Thomas Frazier Rumbold

Pruritic Rhinitis

Its medical and surgical treatment

Thomas Frazier Rumbold

Pruritic Rhinitis
Its medical and surgical treatment

ISBN/EAN: 9783337373429

Printed in Europe, USA, Canada, Australia, Japan

Cover: Foto ©berggeist007 / pixelio.de

More available books at **www.hansebooks.com**

(HAY-FEVER, AUTUMNAL CATARRH, ETC.)

ITS

MEDICAL AND SURGICAL TREATMENT:

WITH EIGHT ILLUSTRATIONS.

BY

THO͏S. ͏͏MBOLD, M. D.,

of the Ameri͏ ͏ociation ; Member of the St. Louis
͏dical Society ; ͏ ͏f the American Medical Association, ͏͏ ͏Medical Association, etc.

MED͏ ͏HING COMPANY.
͏on Avenue.
͏8.

ILLUSTRATIONS.

PREFACE.

The subject of this little monograph is not new, yet but comparatively little has been written on it until the last two or three years. Within this short period, rapid progress has been made in the methods of treatment. The experience of the last two or three years especially, has abundantly proven that what was at one time considered as an irremediable complaint is now under complete control. The methods of investigation formerly employed were barren of results, with the exception of demonstrating that the names given to the ailment were misleading and not descriptive as they purport.

The name I have suggested, PRURIFIC RHINITIS or PRURITUS RHINITIS CATARRHUS or PRURITIC CATARRH, is descriptive of its most prominent and constant characteristics, namely, itching, inflammation and flow of mucus.

I have no doubt the profession will observe—as soon as their attention is called to it—a sufficient number of facts to convince them that this complaint

IS ONE OF THE VERY MANY SEQUENCES OF COMMON NASAL
CATARRH. This I said in a paper read before the St.
Louis Medical Society in May, 1869. I made the
same statement during a discussion on so-called hay-
fever at the Illinois State Medical Society, at its
meeting in Jacksonville, Ill., in May, 1874. On page
60 of my work on HYGIENE AND TREATMENT OF
CHRONIC NASAL CATARRH, is seen the same assertion. Although this work was not given as a whole,
to the public, until January, 1881, yet I gave a few
forms (about 166 pages), in which this statement occurs,
to quite a number of the members of the American
Laryngological Association in June, 1880, and sent
out several hundred copies of the same pages to physicians in the West, as an advertisement of my book.

The first rational step toward the cure of this
complaint, is its treatment as a sequence of chronic
inflammation of the nasal passages. Unless such a
course is persued successfully, even the results of a
radical surgical procedure will only be temporary, in
alleviating the pruritic symptoms. The removal
of a hyperæsthetic membrane CAN NOT arrest the inflammatory process that was the producing cause of
the hyperæsthesia; this is self-evident, consequently
the ultimate recovery of the patient will certainly
depend upon hygienic and constitutional measures
and the spray producers. I am very sure that the
profession will arrive at this conclusion, after they
have had a few additional years experience with this
peculiar complaint.

The manuscript of this little book has been on my table for several years. The reason for the delay in its publication, was owing to the fact that within a few years, a surgical operation was proposed for the immediate relief of the most prominent symptoms of this complaint, and I desired to test whether the claims of its efficacy were well founded.

These claims were put forth by Dr. Wm. H. Daly of Pittsburgh, Pa., and Dr. J. O. Roe, of Rochester, N. Y., who, I believe, were the first to demonstrate to the profession the very great value of this procedure. A part of their views and experience is given in an appendix, with their knowledge and consent.

In this appendix I have also given the views and experience of Dr. P. W. Logan of Knoxville Tenn., and Dr. J. A. Stucky of Lexington, Ky. I was compelled to present their valuable contributions in this form, as the part of my book containing these subjects had been electrotyped before their papers were received.

While freely acknowledging that surgical interference will cut short the pruritic nasal symptoms, yet, I as freely say, I fear many patients will be operated upon, who ought to be cured without the formation of cicatricial tissue in the nasal passages, by being treated for the originating disease, the chronic catarrhal inflammation of the nasal cavities. That this latter course is successful, I know from experience.

THOS. F. RUMBOLD.

St. Louis, June, 1885.

CONTENTS.

CHAPTER V.

CHAPTER VI.

CHAPTER VII.

CHAPTER VIII.

CHAPTER IX.

CHAPTER X.

CHAPTER XI.

APPENDIX.

PRURITUS RHINITIS CATARRHUS;

(HAY-FEVER, SUMMER CATARRH,
AUTUMNAL CATARRH, ETC.)

CHAPTER I.

INTRODUCTION.

ONLY LATELY ATTRACTING THE ATTENTION OF THE PROFESSION.

This complaint, popularly known as "hay-fever", "rose-fever," "June cold," "Autumnal-catarrh",etc, has only of late years attracted the attention of the medical profession. I have not the least hesitancy in saying that this is due to the fact that they have only of late found the department of Rhinology one of interest, or rather one of more importance than was accredited to it a few years ago.

It is a malady that has several peculiar characteristics, among which may be named: its re-occurring, almost uniformly at certain seasons of the year and affecting its victims to the severest degree in certain regions of the country, while at other seasons and in other regions, the great majority enjoy almost complete exemption from its attacks. It has another disagreeable peculiarity, namely, it has not until very lately been relieved by any methods of medical treatment. Indeed so completly have the profession failed to even ameliorate the complaint, that the victims have given up all hope from this quarter and in 1874 formed themselves into a society solely for the purpose of mutually searching for relief. They have agreed to report at any time "during their natural life and afterward if permitted" any remedy for their ailment. Up to last September no such remedy has been reported by their secretary. Such extraordinary measures were never before taken by any class of invalids.

As a matter of interest I will give their constitution in full:

"Article, I. This organization shall be called 'The United States Hay-Fever Association.'

"Article, II. Its object shall be mutual benefit, and seeking for information which shall serve to relcive all sufferers with hay-fever, wherever found.

"Article, III. Any person afflicted with hay-fever or rose-cold may become a member of this association by signing the constitution.

"Article, IV. The officers of this association shall consist of a president, vice-presidents, an advisory board, a treasurer, corresponding secretary, and recording secretary.

"Article, V. It shall be the duty of each member to report to the recording secretary the discovery of any remedy, source of relief, or exempt district which may come to his or her knowledge during their natural life, and afterward, if permitted.

"Article, VI. The secretary, on receipt of such

information, shall apprise each member of the association at their last and usual place of abode.

"ARTICLE, VII. Honorary members may be elected at any meeting of the association.

"ARTICLE, VIII. The annual meeting of the association for the choice of officers and other business shall be held at Bethlehem, N. H. on the last Monday in August in each year, at 4 P. M. Other meetings may be held September, upon the call of any six members of the board of government.

"ARTICLE, IX. This constitution may be amended at any annual meeting by vote of the majority present."

The formation of this organization proves that the despair of ever receiving relief from the medical profession must have been great indeed, and it is a standing condemnation of that profession. Evidently this medical inability—to give it no harsher name—is due to the fact that they have heretofore wholly ignored the study of the diseases affecting the nasal passages and the cavities connected with them.

Until a systematic course of study of this region is pursued and equally systematic investigations are made by a large number of the profession, the successful treatment of all the ailments sequent to catarrhal disease of the nasal cavities will not be made known.

USUAL METHODS OF INVESTIGATION DEFECTIVE.

The methods of investigation usually persued by those who have devoted some time to the study of this complaint have been to receive from the sufferers their own account of their symptoms or condition, instead of endeavoring to ascertain the causes that prepared the victims mucous membrane for the attack.

The answers received related to the dates of the attack and disappearance, and a number of other peculiarities, all of which did more to confuse and mystify than to elucidate the subject.

The investigators have thus laid themselves liable to be led as far astray by these histories, as they would be from the answer of an individual who had a disease, the existence of which he had no sensible knowledge; for instance: the answers of one afflicted with a monomania, if questioned on the subject of his mental ailment. On all other subjects he could, in all probability give correct answers. So with the sufferers of this complaint; on many other matters connected with their disease, except as to the condition of their mucous membrane, they could give correct answers, but on this subject they are VERY LIABLE to give incorrect answer, simply because there are NO SUBJECTIVE SYMPTOMS connected with this peculiar conditon of their nasal passages, and these are the ONLY symptioms they could give.

THE INFLAMMATION THAT PREPARES THE PATIENTS NASAL PASSAGES FOR THIS COMPLAINT CAUSES NO PAIN.

It is because of the non-subjective character of proliferative inflammation, that a large per centage of these sufferers state positively that they were in a perfectly healthy condition up to the period of their first attack and between their attacks. This shows the fallacy of allowing the victims to write the histories of their own complaint, especially when the disease is to be studied from these histories.

This subject will receive further consideration in another portion of this work, but enough has been given to show plainly that this method of investigation is exceedingly liable to be misleading.

I have purposely abstained from considering purely theoretical points, such as accounting for the attacks so frequently occurring in summer and in certain regions of the country, while during cold weather and in a few parts of the country the victims enjoy comparative exemption.

These and other apparently inexplicable features of this neurotic rhinitis, may ultimately assist in the further elucidation of its etiology, which at the present writing is considered unknown.

CHAPTER II.

PRURITUS RHINITIS CATARRHUS ; OR PRURITIC RHINI-
TIS ; OR PRURITIC CATARRH ; OR ITCHING NASAL
CATARRH, (*Hay-Fever, Rose-Cold, June-Cold,
July Cold, Pollen-Fever, Autumnal-Catarrh,
Ragweed-Fever, Peach-Cold, Summer-
Catarrh, etc.*)*

THE PRESENT NAMES INAPPROPRIATE ; A NEW NAME SUGGESTED.

All the names by which this peculiarly phenomenal complaint is known are inappropriate and misleading.

As the medical profession should not agree to a
change of a name of a disease without good and
sufficient reason, I ask a part of your valuable
time, while I endeavor to show from the characteristic symptoms of the complaint that the names given
are misleading, that we have a more appropriate
name in PRURITUS RHINITIS CATARRHUS, OR PRURI-
TIC CATARRH, because the prominent symptom of this
kind of nasal catarrh is an itching.

It is noticeable that in giving names to diseases, it
is oftentimes designed to indicate their nature, by se-

---

* Read before the St. Louis Medical Society, May 3rd,
1884.

lecting some prominent symptom or peculiarity of
the complaint by which to designate it. When such
names are sufficiently descriptive, we may not do
better than to name a disease in this way. Thus,
some names point to the part of the body affected, as
cerebro-spinal meningitis, pneumonitis, rhinitis, laryn-
gitis, otitis, etc; some, the appearance of the patient
while sick, as yellow fever, scarlet fever, spotted fever,
some, the supposed cause of the ailment, as mala-
rial fever, bilious fever, hay-fever; and still others, the
time of the year in which attacks occurs, as summer
catarrh, autumnal fever etc. If all of such names
truly indicated what they seem to do, then they might
very properly be retained, but if any of them indicate
that a certain prominent fact or feature of a disease
is constantly present, so as to distinguish it from
other diseases, when such is not the case, then, most
certainly, the misguiding name should be discarded;
as its retention will be very liable to lead to an erroneous diagnosis, and thus a case might be excluded
from its proper class and, as a consequence, be improperly treated.

Recent investigations, of a very thorough character, go to prove that the various names by which
this complaint is designated, are misleading. I am
not unmindful of the fact that Dr. Morrill Wyman, a
high authority on this complaint, regards the spring
and fall forms as separate diseases. In this, I think
that he is mistaken. On this point I agree with Dr.
G. M. Beard, also an excellent authority on "hay-

fever." Dr. Beard says: "In view of the large number of facts afterward obtained, and which are recorded in this work, it was found necessary to aban-
don this theory [the two forms of the complaint]
and to admit the substantial identity of 'Autumnal
Catarrh' and 'June Cold'." If the dates of attack
and disappearance were erased from the history of a
case of this disease, I defy, even an expert, to do
better than guess the season of the year in which it
occurred, nor could he make a better guess as to the
duration of the attack. I am unable, after a very
careful study of Dr. Wyman's really valuable work,
to perceive the difference between the spring and
fall forms of the complaint, except one of severity
and that they almost uniformly occur, the one in the
spring and the other in the fall; but the individuals
who have the attack uniformly in May or in June
or in July, relate, *identically*, the same symptoms as do
those whose date of attack occurs in August or September, the fall form being the more severe. All
kinds of this class of sufferers are exempt from attack
by resorting to the same mountainous regions of the
country and, according to my experience, all are relieved by the same kind of hygienic management and
the same kind of constitutional and local treatment.
Cases are not at all uncommon who may, for a few
years, be afflicted in the early summer months but,
from some unknown cause, pass the usual period of
attack, and experience it either later or earlier than
usual. I have a young patient under treatment now,

whose first attack occurred in July, the next in May and the next in September.

That others consider the name not the most suitable, is seen from the following quotation taken from Dr. Beard's work. He says:

"The inappropriateness or rather the insufficiency of the term hay-fever is now quite generally admitted; for even where the predisposition exists, hay of any kind, fresh or dried, acts as an exciting cause in but a minority of cases, and rarely, if over, is it the only irritant that gives rise to the paroxysms."

The name "hay-fever" indicates that hay alone is the cause of the attack, which is very far from being the case. I have a patient who can handle hay at any period of the year without experiencing the least inconvenience, another one who is not the least affected by it, so long as his scalp and face are not moist with perspiration. While this patient is perspiring during warm months, any kind of dust, but especially that from an old carpet, instantly sets him wild with an itching sensation of the face and eyes, soon followed by the same sensation in the nostrils and by sneezing.

The same objection exists with respect to the names "rose-cold," "pollen-fever," etc. It is almost universally admitted that any one kind of pollen, or any one kind of flower, may seriously affect some persons, and have no bad effects on others, yet the distinguishing phenomena, namely the itching, flow of tears, the flow of watery secretion from the nostrils are nearly alike in all patients, whether they are attacked in the spring, summer, fall or winter. If they differ, it is in degree of severity only.

This brings us to the names which designate the seasons of the year in which the disease occurs. If the seizures uniformly commenced in June, July or during the autumnal months, the name of the month or of the season of the year might very properly be prefixed to the word "cold" or "catarrh" or "fever" or "asthma," but my observations since 1862, and the very thorough investigations of Dr. Beard, leave no doubt that the attacks may occur in any month during the summer. Because of its so frequent appearance during the summer months, Dr. John Bostock, of London, [1819] suggested the name "Catarrhus Aestivus" or "Summer Catarrh." This also is misleading, as well as Dr. M. Wyman's name, "Catarrhus Autumnalis" or "Autumnal Catarrh." These names indicate that individuals could not be attacked during cold weather; but it is *well* known that the complaint may sometimes affect its victims as late as October, November, December and even January, according to Dr. Beard's report. I had a patient who had attacks in every month from April to November, and I have one now (May, 1884) under treatment who has had attacks for two years; and the whole year around whenever he is where the air is hot and dusty. (I have three patients now under treatment who have had severe attacks in the last week in January, 1885.)

It does not detract from the value of these facts,

to say that these last patients and all other like patients who have been under my care, had their winter attacks much less severely than their warm weather attacks, nor is the argument weakened by the fact that the very great majority of attacks of this complaint occur in warm weather. As the *very same* symptoms occur in cold weather, warm-weather names are misleading.

The following tables will give, in a condensed form, the dates of attack and of disappearance. These tables, which are taken from Dr. Beard's valuable work, do not show the duration of the attack. He received his information from answers to inquiries sent to individuals afflicted with this complaint, numbering 200.

TABLE OF DATES OF ATTACK.

From May 1 to 10, 2.	From Aug. 1 to 10, 7.
" " 10 to 31, 6.	" " 10 to 20, 81.
" June 1 to 10, 11.	" " 20 to 31, 54.
" " 10 to 80, 8.	" Sept. 1 to 10, 7.
" July 1 to 10, 6.	" " 10 to 20, 1.
	" " 20 to 30, 2.
" " 10 to 20, 6.	" June to Sept. 1.
" " 20 to 81, 7.	" Aug. to Jan. 1.

As to dates of disappearance the answers received were the following:

TABLE OF DATES OF DISAPPEARANCE.

January or early winter. - - - -	2.
About January 1st. - - - - -	1.
Late in winter. - - - - - -	1.
March 1st. - - - - - - -	1.
Middle of July. - - - - - -	6.
Latter part of July. . - - - - -	5.

Early in August. - - - - - 5.
Middle of August. - - - - - 2.
Latter part of August. - - - - 1.
Early in September. - - - - - 2.
Middle of September. - - - - 13.
Latter part of September. - - - - 26.
Early in October. - - - - - 42.
Middle part of October. - - - - 14.
Latter part of October. - - - - 8.
Early in November. - - - - - 9.
Middle of November. - - - - 4.
Early in December. - - - - - 1.
Middle of December. - - - - 1.
From September 15 to December 25. - - 1.
With frost or cold weather. - - - 87.
Three weeks after beginning. - - - 1.
Cannot state definitely. - - - - 1.

It is self-evident, from the facts shown by these tables, that "Rose-Cold," "June-Cold," "Hay-Fever," "July-Cold," "Pollen-Fever" "Summer-Catarrh," "Autumnal-Catarrh," are all inappropriate, or rather insufficient names, and that any one of them tends to mislead the physician who would allow himself to be guided by the characteristic suggested by the name.

A strong point in favor of the parasitic or vegetable theory is made in the constancy and regularity of the appearance of the disease at given times with *some* of the victims, not only coming on at a fixed day, but the very hour, and also its almost regular disappearance at such times as might usually be expected that the spores of the bacteria or vegetable growths would be destroyed by natural causes. If this nasal trouble is caused by germs, why may not other nasal troubles originate from germs? Without answering this question I will ask another: How could germs cause this trouble when they depend upon a peculiar condition of the fluids of the mucous membrane for sustenance, which condition must have been the result of diseased action? As these germs do not in this way affect a healthy mucous membrane, does this not show that diseased action was primary, and germ irritation secondary? Dr. Beard while speaking on the vegetable theory, says: "This suggests to almost any one the possibility that some parasitic or vegetable emanation appearing only during the season of the disease might be the cause.

"If it could be shown that some at least of the symptoms were felt at other than these so-called catarrhal seasons; if sufficient evidence of the occurrence of certain phases of the malady in the winter and spring could be obtained, the parasitic and vegetable theories would be seriously shaken.

"This evidence is here given. The hay-fever symptoms that are in the winter excited by exposure to the dust of hay or of the house, or to animal emanations, are usually, if not always, of a transient character, lasting but a few minutes or hours; but for this brief time they are characteristic of the disease, and they do not appear in other persons."

He arrives at these conculsions from answers to the following question: "Do you ever have, during the winter or spring, when exposed to any of the exciting causes, as dust etc., attacks resembling 'hay-fever' in a mild form, lasting perhaps for a few minutes or hours?"

Of 200 affected individuals, 101 answered Yes, 77, No. His special replies to the same question, contain these significant expressions: "Lots of e'm, but mild in form," "For a few hours," "Dust of hay will cause it." "Caused by dust," "Dust of sweeping, etc."

Pruritus Rhinitis Cata rrhus, or Pruritic Rhinitis, or *Pruritic Nasal Catarrh, or Itching Nasal Catarrh*, is the name that I have selected for this phenomenal complaint. This name is descriptive of its most prominent and constant characteristics, namely *itching, inflammation and flow of mucus.*

The attack is ushered in by an itching of the nose and face; this soon affects the eyes, causing intense suffering. The itching sensation in the nostrils gives rise to prolonged sneezing; this, in turn, makes the eyes still worse; presently, the itching reaches the soft palate and the fauces, and to relieve these parts of this same sensation, the tongue is used to rub them. As the tickling is not relieved a rasping cough is tried, which is so persistently continued that the throat soon becomes sore, and in older sufferers, shortness of breath ensues, and symptoms of asthma are developed. I have not had a patient that did not experience this itching early in the disease, and it was always prominent. Dr. Beard, gives this as the first symptom and says of it, on page 118. " This is one of the first, oftentimes the very first local symptom of an attack." Dr. Wyman, in his work on "Autumnal Catarrh" in mentioning the local symptoms as they occur consecutively, says, on page

12. "The lining membrane of the nostrils is the part first affected : beginning with a slight tickling or itching, which soon shoots upwards towards the eyes, and even into them."

To repeat, because of the uniformity of this symptom, and the fact that it is always accompanied by inflammation, I think the name PRURITUS RHINITIS CATARRHUS or PRURITIC RHINITIS or *Pruritic Nasal Catarrh or Itching Nasal Catarrh* is more descriptive of the complaint than any of the names now given to it. This name indicates the first, the principal and the most prominent symptom and which is truly characistic of the malady at whatever season of the year the victim is attacked, and it is not misleading.

CHAPTER III.

Little or no benefit is to be derived from the study of the history of the literature of this complaint. For this reason, only the outline will be given, and that as conscisely as possible.

In 1819 Dr. John Bostock of London presented a paper to the London Medico-Chirurgical Society, in which he gave the first formal description of this complaint, describing his own case. The title of his paper was a "Case of a Periodical Affection of the Eyes and Chest." This paper was published in the Medico-Chirurgical Transactions, page 161, part 1. volume x. In 1824 he gave the complaint the name of "Hay-Fever".

In 1828 the same author read another paper on the same subject and gave the disease the name of Catarrhus Æstivus or Summer Catarrh. This also was published in the same Transactions, Volume XII, page 437. This name is still retained by many writers.

In this year (1828), Dr. Mac Culloch published "An Essay on Remittent and Intermittent Diseases". In this he mentions a complaint that he thought was caused by emanations from hot-houses and greenhouses, but especially from hay-fields. He says that the "common people observed that the disease was brought on by exposure during hay-making seasons."

In 1829 Mr. W. Gordon published a paper in the London Medical Gazette, Volume IV, page 266, on the "Observations on the Nature, Cause and Treatment of Hay-Asthma". He thought that the flowers of grass was the cause. For this reason he thought the complaint should be called "grass-asthma" instead of "hay-asthma".

In 1830 Mr. A. Praster published the history of a case.

In 1831 and in 1833 Dr. Elliotson referred to a complaint resembling this one, and published it in the London Medical Gazette. He rejects the heat theory of Dr. John Bostock, and the hay theory also, and affirms that grass and probably the pollen of flowers are the causes.

In 1847 Dr. Ramadge, in his work on "Asthma," published in London, holds that the emanations of of grass and flowers give rise to attacks of this disease.

In 1850 Dr. Gream published a paper in the London Lancet, Volume I. page 692, on the "Use of Nux Vomica as a Remedy in Hay-Fever." In this paper he affirms that neither the flowers of grasses or any other flowers are the producing causes, but that it is in-door and out-door dust that are the exciting causes. He observed that after a rain the victims were much relieved.

In 1852 Dr. La Forgue of Toulouse, wrote a paper on this complaint. He advocated the heat theory.

In 1854 Dr. Morrell Wyman of Cambridge. Mass. being himself a sufferer of what he terms " Autumnal Catarrh," discribed the complaint in his course of lectures in the Medical School of Harvard University. He says " the description was drawn from my personal experience, and a few cases which had come under my observation for treatment."

In 1857 Dr. Watson says that this malady is caused by vegetable emanations floating in the air.

In 1859 Dr. Walsh in his " Treatise on the Diseases of the Lungs," refers to this complaint and calls it a singular variety of " naso-pulmonary catarrh." He says that the complaint occurs during hay-making time and that the odor from this and from grasses are the causes, but he states that he has the history of a well attested case in which the victim suffered from an attack while on a voyage across the Atlantic. Each day's symptoms were given.

In this year (1859), Dr. H. Salter, the asthmatic author of a work on " Asthma " states the complaint is a hay-asthma and lasts during the hay-making season, and adds, that heat, dust and sun shine are the agencies that most frequently excite an attack.

Dr. Philip Phœbus, a professor in the University of Giessen wrote in this year also, far out stripping his predecessors in the thoroughness of his investigations.

From extensive correspondence, which was exten-

ded almost through the entire civilized world, he
gathered many facts not before known. But he was
not satisfied with his conclusions as to the etiology of
the complaint; he says that it will require more accurate observation to come to positive conclusions.
He thought that odors of flowers, probably that of
rye was the most irritating, and the irritation of dust
assisted by long heat and ozone were the causes of the
attacks. His investigations were published in 1862.

In 1860 Dr. Koranz of Neufchatel, Switzerland,
wrote a paper on "Hay-Fever," and published it in
the Le' Echo Medical. He thought that the flowers
of grass was the irritating cause.

In May 1866 the facts known to Dr. Morrell Wyman
were embodied in a paper read at the Annual meeting
of the Massachusetts Medical Society in Boston. The
following abstract was published in the Boston Journal June 2nd, 1866.

"Autumnal Catarrh : At a meeting of the Massachusetts Medical Society, Dr. Wyman of Cambridge
gave an account of a singular catarrhal affection, or
cold, hitherto undescribed, and named by him Autumnal Catarrh. There are two annually appearing
catarrhs in this country : the summer catarrh (com-
monly called Rose Cold, Hay Fever or June Cold)
begins the last week in May or the first week in June,
and lasts four or five weeks ; the other, the Autumnal
Catarrh, commences the last week in August, and continues till the last week in September. It begins with
sneezing, itching of the eyes, especially at the inner
corners, watering of the eyes, and a profuse discharge

from the nostrils. The affection of the eyes is in fits, coming on suddenly, compelling the sufferer to rub his eyes violently for relief. The fits of sneezing and nose-blowing and obstruction of the nostrils are also sudden, and when the fits are over, usually in a few minutes, go off as quick as they come. During the second week in September, a cough sets in, dry, violent, and in fits; it is increased during dry, dusty weather, and relieved by an easterly storm. It is severe in the night, and there is sometimes asthma. The disease subsides during the third week in September, and by the first of October, or the first good frost, is entirely gone. It is not an uncommon disease; Daniel Webester had it annually for twenty years, and while Secretary of State suggested to President Fillmore the propriety of resigning on account of it. The late Chief Justice Shaw of Massachusetts was another victim. Medicines seem to have been most freely tried without materially relieving its severity or shortening its duration. Fortunately, it has been ascertained that there is a most complete and agreeable cure.* Within twenty-four hours after the sufferer arrives at the White Mountains, at Gorham, at the Glen House, or the Waumbec, it suddenly disappears and if he remains till the last week in September, the usual time of disappearance, does not return for that year. The

---

* I am astonished that such an accurate man as Dr. Wyman, should employ the word " cure " in this sentence, when he means complete cessation of the symptoms during that catarrhal season, on condition that the victim remains in the locality named. Cure, means complete, continuous recovery without recurrence at any season, complete return to health. He does not desire that this should be inferred.

relief at Franconia is not so certain as at the north side of the mountains, though most are relieved there also."

In this year (1866) Dr. W. A. Smith also published a work " on Hay-Fever, Hay-Asthma or Summer Catarrh." He thought the worst symptoms were brought on by great heat, grass and flowers. With great fairness he gives the history of a patient that proves that his views are not invariably correct; namely:

" This year the disease first came on while I was on the sea yachting with a friend. It was a hot day in May, with wind from the southwest, the nearest land to windward being nine miles distant. I felt myself, after some exertion in assisting to hoist the sails, suddenly seized with sneezing, and I have had it ever since."*That is up to June 13th, of that year.

In 1867 Dr. W. Pirrie of London, published a work on " Hay-Asthma and the affection termed Hay-Fever." He added little to the literature of the subject, except the very important idea that the nervous system was a far more important factor to the complaint than had heretofore been supposed.

" This little treatise of Dr. Pirrie is remarkable that in many respects it theoretically anticipates what by these statistics and facts, has been demonstrated in regard to the nature of hay-fever, and the true principles of treatment as now confirmed by a large induction in Europe and America. The number of his facts was so limited, and the cases he gave were so imperfectly detailed, that none of the points he suggested

---

* Beard on Hay-Fever, 1876.

could be considered as proved; consequently, they
have not been generally received, and have excited
comparatively little attention; they were suggestions
and nothing more, and they left the subject as mysteri-
ous as they found it."

What author has not left the subject as mysterous
as he found it?

In 1869 Prof. Carl. Binz of Bonn, Germany, contributed an article to Virchow's Archives for February,
on the use of quinine as a remedy for this complaint.
In this article he gives a letter addressed to him from
the illustrous Helmholtz recommending the local application of the sulphate of quinine as a remedy.

In this year (1869), in a paper read before the St.
Louis Medical Society on the sequences of chronic
nasal catarrh, I said that careful investigation would
substantiate my assertion that "hay-fever" was one of
the sequences of chronic inflammation of the mucous
membrane of the nasal passages, giving at the same
time the history of the symptoms and the treatment
of the first case given in this work. In two or three
years after this I re-asserted these views before the
Illinois State Medical Society, at its meeting in Jacksonville, on the occasion of a discussion on the effects
of quinine when applied to the nostrils as a cure for
hay-fever.

In 1870 Dr. G. Moore of London, published a work
on "Hay-Fever or Summer Catarrh; its Causes, Symptoms, Prevention and Treatment." He favored the
theory of sunlight, heat and the effluvia of hay and

flowers as will as decomposing vegetable matter.

In 1872 Dr. Morrell Wyman again contributed his experience and observations of "Autumnal Catarrh (Hay-Fever.)"

"A leading thought in this work of Dr. Wyman is that in the United States, under the general term hay-fever, two distinct forms of disease are included the so called 'Rose-Cold' or 'June Cold' occurring in May or June, and corresponding to the 'hay-fever' or 'hay-asthma' of England and the continent, and a later form, beginning in August and lasting several weeks in the fall, to which he gave the name 'Autumnal Catarrh.' "*

In 1873, Mr. Chas. H. Blacklay an eminent surgeon of Manchester, England published a work on "Experimental Researches on the Cause and Nature of Catarrhus Æstivus (Hay-Fever or Hay-Asthma)." He advocates that the pollen of grass is almost exclusively, the cause of the attacks of the complaint, and thinks that the names " Pollen-Fever" or " Pollen Catarrh " more appropriate. He made a large number of experiments with the fresh and dried pollen of the grasses and of other plants and believes that he substantiated his theory. He says that rabbits, guinea-pigs and cats are similiarly affected by the pollen.

In June 1876, the late Dr. George M. Beard published an excellent work on this subject, and claims to have discovered, what he terms a " middle form of hay-fever." He advocates what he calls the nerve theory, and, according to my views, is far in advance of all

---

* Beard on Hay-Fever.

previous authors. The following quotation from the
preface of his work, will give a partial idea of his
views:

"The theory of this book, that this disease is a com-
plex resultant of a nervous system especially sensative in this direction, acted upon by the enervating influence of heat, and by any one or several of a large
number of vegetable and other irritants, has the advantage over other theories; that it accounts for all
the phenomena exhibited by the disease in this or in
any other country."

In August of this same year we are given a most
carefully prepared work on " Autumnal Catarrh (Hay-
Fever)" with illustrative maps by Morrell Wyman, M.
D., of Cambridge Mass. Every physician who desires
to study this complaint should supply himself with
this work, as well as with Dr. Beard's.

This last Edition of Dr. Wyman contains his former
views amplified. He holds that his Autumnal Catarrh
is a different complaint from that of the early summer catarrhs, i. e. the " Rose or June Cold."

As I will take the liberty to make frequent quotations from Dr. Beard's and Dr. Wyman's valuable
works, I will not, at present, give more of their views.

Of course I shall differ radically from both of these
talented gentlemen, but wish to acknowledge that I
have received more information, in the study of this
complaint, from their works, than from all the various works on this subject that has come under my
notice which number in all about fifteen.

CHAPTER IV.

Pruritus Rhinitis Catarrhus (*Hay-fever, Summer Catarrh, Autumnal Catarrh, etc.*) ONE OF THE SEUQENCES OF CHRONIC CATARRHAL INFLAMMATION OF THE NASAL CAVITIES. *

Authors have had a suspicion for many years, that this disease ("hay-fever") might, in some way, be connected with common nasal catarrh; consequently, they all have given this part of the subject some attention, but for various reasons they have come to the conclusion that there is no evidence of such relationship.

It seems to me that their methods of carrying

*Read before the St. Louis Medical Society May 10th, 1884.

on their investigations, concerning this matter, have not been thorough enough, and with some of the authors, they have been quite defective.

They have asked sufferers, all of whom, with a very few exceptions, resided at a distance, questions, the design or tendency of which they could not fully understand, not being medically educated. In fact they have taken it for granted, that these individuals knew the cause and course of their malady, and the questions have been so formed, that when filled out they completed the histories of just such cases, as the victims and the authors had conceived them to be.

It always requires much greater medical acumen to make a diagnosis, than it does to write a prescription for a known disease. Now, while no medical man has been known to ask an ailing individual to write his own prescription, yet the authors on pruritic catarrh have asked their correspondents to take the more difficult part, namely, the writing out their own diagnosis, and from these papers they have studied the complaint; and what makes these narrations of still less value, they are, almost universally dated from a period *after* their first most chracteristic attack, and not from their initiatory symptoms. These, the sufferers would not recall, unless assisted by interrogations conducted by one acquainted with the peculiarities of such cases. To say the least, this is a defective method, especally when there exists a supposition that the complaint might be secondary to another disease. Under these circumstances, why not, make enquiry concerning their physicial condition previous to the first attack of pruritic rhinitis (hay-fever). Without this, their investigations are illogical, as they have left their readers ignorant of the condition of the system that might have made the attack possible, if the nasal inflammation preceded it or caused it.

I am fully aware that my views on this subject are not in accord with any of the authorities, and in taking this position, it devolves upon me to prove that this phenomenal complaint is a sequence of comparatively long standing inflammation of the mucous membrame of the nasal cavites. This I will do by giving accurate and detailed histories of the physicial condition of those who have been my patients, which will show that the inflammation always precedes it. It is evident that this will go far to sustain my proposition; but to make it still stronger, and because some might say that the co-existence of long continued inflammation in the nasal cavities was a co-incident and would not, of itself, necessarily prove that pruritic catarrh, was occasioned by it, I will give other evidences that will demonstrate beyond the possibility of a doubt, the relationship of the two complaints. This will be done by giving the histories of patients whose ameliorative treatment for chronic nasal inflammation, reduced the frequency and the severity of their attacks of pruritic catarrh and of a few other patients whose treatment caused an entire cessation of the disórder.

In 1869, I made a statment before this society, that a scrutinizing investigation of the patient's condition, during that period previous to the first attack, would show that chronic nasal inflammation had rendered them liable to be afflicted with pruritic catarrh. My numerous observations, made since that date confirm me in this matter. In fact, every individual, whether patient or acquaintance, that I have seen since 1862, who had suffered from attacks of it, had been for several years afflicted by chronic catarrhal inflammation of the nasal cavities.

I am not prepared, at present, to give my reasons for this neurotic form of rhinitis attacking the great majority of its victims in summer days and in certain regions of the country, while during cold weather, and in a few parts of the country they enjoy comparative exémption. These and other apparently inexplicable peculiarities, may ultimately assist in the further elucidation of its etiology, which at present writing is considered unknown. Before giving these clinical facts, I wish to show how both the investigators of this complaint and the sufferers interrogated have made grave mistakes.

Upon the occurrence of an inflammation of the mucous membrane, the blood vessels are not only filled to their utmost capacity, but they are greatly enlarged by reason of their excessive engorgement, being increased from 10 to 40 times their normal

diameter, according to the severity of the irritation.
If this inflammation should become continuous by
repeated irritations for a number of years, the excessive amount of blood (nutrition) going to the part,
causes its permanent thickening, just as an inflamed
and enlarged joint will be permanently enlarged, if
the inflammation shall be allowed to continue for a
long time. In the case of the mucous membrane, this
growth is denominated proliferative inflammation.
It is during this stage of the inflammatory disease of
the nasal passages, that the patient may, from some
cause at present not known to the profession become
affected with pruritic rhinitis.

According to my observation, a most important characteristic of proliferative inflammation (and it is one
that should be continually borne in mind), is that the
patient does not experience the least sensation of
pain during its progress.   Not until the caliber of
the air spaces in the nostrils are so reduced in size,
that respiration is thereby impeded, do they experience the least inconvenience, except it may be that
they have slowly, imperceptibly lost the sense of
smell from the same cause ; or, this abnormal process
may stealthily invade the Eustachian tubes and middle ears, and slowly and imperceptibly rob its victim
of his hearing, but if the loss of these senses should not
suggest the presence of this inflammatory process, he
would be entirely unconscious of it, so perfectly painless is its growth.

Another dangerous peculiarity of this variety of

inflammation is, that the patient rarely experiences the usual well known symptoms of "catching cold," or at least a very severe cold, yet the proliferative process, that is the abnormal change of the mucous membrane, is continuous.

It is evident that, with this state of things, it is impossible for Dr. Beard's or Dr. Wyman's correspondents, to have had the least idea that they were victims of this variety of inflammation, the very kind, the only kind that could prepare their nasal mucous membrane for the development of neurotic symptoms. When these physicians did not observe this condition, is to be expected that the patients could have made mention of it when not conscious of its existence?

STATEMENTS TAKEN FROM THE EARLY HISTORIES OF PATIENTS SUFFERING FROM PRURITIC RHINITIS, OR ITCHING NASAL CATARRH, TO PROVE THAT IT IS A SEQUENCE OF CHRONIC CATARRHAL INFLAMMATION OF THE NASAL PASSAGES.

I will not attempt to give lengthy details of the early history of each patient, nor an exhaustive statement of the symptoms when he first visited me. Of the early history, I will give that much only that is required to prove that chronic inflammation of the nasal passages always precedes the attack of pruritic rhinitis or itching nasal catarrh, and give the dates, treatment and the result. The plan of treatment will follow at another time.

Physicians will be surprised at the uniformity wtih which the majority of these patients state, at their first visit, that they have been in usual good health previous to this first attack of itching catarrh, and also at the shortness of their memory concerning their symptoms for even a few days or weeks past, but if assisted by various questions, somewhat leading in their character, they will be enabled to recall a sufficient number of incidents that make the history quite complete, which will be amplified by future conversations at subsequent visits during their treatment.

The first case that I will report was, in this respect, a very decided exception as it was during my conversation with him that I was made certain that my views were correct concerning the relationship of this complaint to chronic nasal catarrh.

CASE I. Mr. Luke R. Gibson, æt 43 years, a printer, visited me on June 10th, 1867, desiring relief from his attacks of sneezing and asthma. These sudden attacks commenced in July, 1865. The next commenced in July, 1866 ; on this occasion it occurred on a hot night about the middle of the month, immediately after he had left the printing office, between three and four o'clock in the morning. He thought the exposure to the night air was the cause of the attack.

He voluntarily said that he believed that his chronic catarrh, which he had had since he could remember, was "the cause of the sneezing spells."

EARLY HISTORY.

When a boy he had large crusts of secretion form

in both nostrils. As he grew older these disappeared, but with their disappearance he began to be affected with severe headaches, especially over his forehead. Both of his ears were diseased, and he had had an otorrhœal discharge since his boyhood.

His first attack of itching of his face and eyes commenced one hot morning in July, 1865, as he left the printing office. He noticed at the same time that his usual catarrh had abated to a marked degree, and that as his sneezing grew less, which was about September, his catarrh recommenced. This has always been the case with these two complaints.

My attempt, at the time, to alleviate his sufferings, was productive of positive harm. He visited me on Monday, June 24, at which time the above history was given me. At this time I took two aural polypi from his left ear.

Jan. 4th, 1868, he again visited me. His catarrh was very bad, and he had severe headaches. For this he was treated about three times a week until Feb. 3d, then two times each week until the 26th. After the treatment on this day, he had a slight attack of the itching of the face and eyes, but he did not sneeze. My treatment at this time was too irritating, I was applying by the spray producer a mixture of muriate of ammonia, tincture of iodine and tincture of aconite root.

He at once went to St. Paul, where he resided until Sept., but was not entirely free of his tormentor.

May 31st, 1869. Treated him two times a week, through June and up to July 23d; after this about once a week until Aug. 21st.

He had no attack up to this date, but sometimes experienced sensations of the beginning of the itching of the eyes and edges of the alæ of the nostrils.

He passed Aug. 1870, without a recurrence of his pruritic rhinitis, but lived most of the time in the country. I have not heard of him since that date.

CASE II.—Dr. R. J. P., 50 years, Dentist of St. Louis, consulted me Dec. 19th, 1868, for severe frontal headache; for this I treated him until March 1st. During his visits he informed me that he was subject to what he and his physician called "hay-catarrh." It usually attacked him in May. I recommended the continuation of the treatment as a preventive. To this he acceded. But as I was not successful in giving him a local treatment, without causing sneezing, he discontinued about the middle of May.

His usual attack commenced this year on May 30th, after taking a Turkish bath on Sunday morning. For relief he immediately visited the mountains of Tennessee, returning to the city in the fall.

March 1st, 1870, I commenced to give him treatment for his catarrh, this was continued for a few weeks, three times a week, then twice a week until the 14th of May, at which time he feared the itching catarrh would attack him.

EARLY HISTORY.

At these visits I learned from him that he had been subject to sore throat, enlarged tonsils and severe headaches, as well as continual clearing of his throat in the mornings, since he was a boy. When his early history was first spoken of, he had forgotten all his early troubles about his throat and head, as these had not troubled him so much of late years, except on the occation of his visits to me.

He started for Louisville, Ky., and arrived May 16th, 1870, and concluded to remain there a few days before going to Tennessee; but this visit was prolonged to the Fourth of July, at which time, as he had missed his "hay-catarrh," he concluded to return to St. Louis. On his way home he was attacked on the cars. He continued on his journey home, remained quiet for a few days and entirely recovered.

I gave him a few treatments in May, 1871. He remained in the city the whole of that year entirely exempt from his "hay-catarrh." I have not heard from him since the fall of that year.

CASE III.—Mr. J. Whaling, of Belleville, Ill., æt 37 years, consulted me May 20th, 1872.

EARLY HISTORY.

During the last ten years has been in much better health than before that time. When 14 or 15 years of age, he had the measels, which left him with a severe cough and diseased ears. For many years after this attack of measels, he suffered from dizziness; would not walk down stairs without taking hold of the hand rail. Has had tinnitus aurium since his ears have been affected.

He now has what he calls "rose-cold," and has had it every spring during the last three years. At first the attacks were not severe, always commencing in May, sometimes the first part and sometimes in the latter part of the month. This time it attacked him after he had taken his supper on Saturday the 18th of May.

I treated him but two weeks with so little benefit, that he left me. since which time I have not heard of him.

CASE IV.—Mr. W. K. G., of Memphis, Tenn., æt 33 years, consulted me Aug. 22d, 1873 for "hayfever."

EARLY HISTORY.

Did not remember of being particularly liable to take cold after he was 21 years old. Up to that age lived a very exposed life. Did not remember when he did not smoke or chew tobacco. Nearly all his life had to clear his throat in the morning, and while endeavoring to do so would become sick at the stomach. If these efforts were made after he had his breakfast, he would throw up his meal.

Treated him but eight times, with little, maybe no benefit. Have not heard from him since.

CASE V.—Mr. Robt. G. Kane, of Alton, Ill, æt 35 years, consulted me Sept. 6th, 1873 for "grass-fever."

EARLY HISTORY.

When a boy always had a running nose, and kept his mouth open. His mother tied a handkerchief over his head and under his chin, so as to make him break his "habit" of breathing with his mouth open. Could not leave it there because it smothered him. This condition of breathing lasted until he was almost a young man, at which time had his "palate" (?)(uvula) clipped off, because of a severe cough. He was at this date taken away from college, because of the cough, and was given cod-liver oil. At *no time* had headache, ear ache or sore throat, nor any kind of a pain, nor was he ever conscious of taking the least cold.

This is the kind of case that both Dr. Beard and Dr. Wyman would say was not liable to "take cold," because the patient says that he has had no cold, therefore there would be no necessary relationship between his "grass-fever" and his very severe chronic nasal catarrh, the proof of the existence of which he had just given, yet he insists that he has never taken a cold in his life, proving that one may take cold, even very frequently, without being conscious of it, as he undoubtedly did.

As I required him to discontinue the use of tobacco,

which he acknowledged was injuring him, he preferred
to go to the Lake Superior regions, where he could
carry on his excesses and still be free from his pruritic catarah.

CASE VI.—Mr. Francis B. A., Hannibal, Mo., æt 34
years consulted me June 16th, 1874.

EARLY HISTORY.

Required the frequent use of his handkerchief when
he was a boy. Was always very small for his age
until he attained his 19th year, then grew up rapidly.
Up to this age they considered him very liable to take
consumption, for which he took cod liver oil for
nearly three years. To this remedy he attributed
his sudden growth and subsequent good health. Had
not been liable to very bad colds, but took slight colds
every winter. This he knew because he experienced
difficulty in singing. At the age of 22 years he acquired the habit of smoking tobacco. In a short time,
may be six months, he observed that he breathed
with difficulty through both nostrils, especially through
the left one, and slept with his mouth open, so that
his throat was very dry and slightly sore every morning. At the same time he had severe coughing spells
in the morning and in his effort to clear his throat
would frequently end in throwing something off his
stomach. As soon as this took place he considered
his cough over for that morning.

The severe sneezing and the weeping of the eyes
commenced last July (1873) while on a train. At
that time, he would put a silk handkerchief over his
nose, as he passed from one passenger car to the other;
in this way he, in a measure, escaped the bad effect
of the wind, the locomotive smoke and the dust. Sometimes on entering the car he would sneeze fifteen or
twenty times before he could attend to his duties as a

conductor. This condition of his case lasted until the first snow. In the early part of this month (June 8, 1874) he experienced the same sensation in an exaggerated form.

Treated him from the 16th to the 21st inclusive once each day, the effect was all that could be desired. His relief was so marked that he thought that the complaint was going away. Being convinced of this, he started home Sunday night the 21st, but he returned on Thursday the 25th, feeling as bad as ever. As he had vexed me by going home so suddenly, and as the appearance of his eyes was far worse than when I first saw him, I, at that time, refused to take his case, but in the evening of that day, I gave him another treatment, the effect of which was not at all encouraging. The fact was, I had made too great an effort to benefit him by the applications, and the result proved that he was too strongly treated, or, in other words, he was over-treated, therefore made worse.

I treated him Friday and Saturday, the effects of which were highly encouraging. I concluded to miss Sunday; this was found on Monday to be an error on my part, as the itching in his eyes returned to a slight degree. He was treated once each day during the next seven days. As he remained in good condition, I concluded to pass the next day, Monday; this was another error in treatment, but the itching was but slight, in fact, scarcely perceptible, but sufficient to determine me to treat him once each day for the next seven days, at which time one days intermission was again tried and found to be sufficient.

He was treated every other day until Aug. 1st, at which time he went home entirely recovered.

He took spray producers with him and continued

the applications at such time as he felt the necessity for them.

I learned that he had passed the next three years without the least return of the complaint. Since that time I have not heard from him.

CASE VII.—Mr. Wm. C. F., Kirkwood, Mo., æt. about 50 years, consulted me May 2nd, 1874.

EARLY HISTORY.

Had enlarged tonsils since he has been ten years old, and frequent abscesses in the throat (tonsils) in the fall months. Has always had tinnitus aurium.

Last Aug. (1873) had severe attacks of itching of the eyes and sneezing whenever he went through a clover field. A week before coming to me he felt the same sensation coming on again. As he was told his complaint was "hay-fever" he concluded to try the effect of treatment. He was quite a large man and had a few attacks of short breathing that resembled asthma.

The first treatment was quite beneficial. I required that he should avoid the clover and hay field, this he did. He was treated the next two days and felt so well that he passed the next day without treatment. May 6th came for treatment, felt well since last visit. Was not treated until the 11th; had a slight sensation of itching and a few sneezes the day before; treated the 17th; then again on the 19th and 23d. As he felt entirely well he concluded that he did not require further medical treatment.

July 25th returned for treatment; had a few sensations of return during the week. Was treated again on the 30th, Aug. 20th, 26th and Sept. 3d.

As he had no evidence of his "pollen-fever" he did not deem it necessary to take further treatment at that time.

He passed the spring of 1875 feeling in unusually good health, but by advice avoided clover and hay fields. Oct. 23d, 1875, came again for treatment, "driven by thirty or forty sneezes that almost killed me yesterday."

He sneezes very strongly, and being a man of about 225 pounds, it tortured him very much. When he sneezed yesterday, he came near falling off his chair, having no control of himself while in the spasm.

Was treated again the 25th, 27th, 29th, Nov. 3d and 10th.

He was treated three times in Aug. 1876, more as a precautionary measure than as a necessity. I have not treated him since; nor have I heard concerning his health, but I have every reason to think that he has remained well, as he still lives in this county.

CASE VIII.—Miss O. H., principal of one our public schools, æt. about 38 years, consulted me March 4th, 1874, because of stoppage of the nostrils, severe frontal headache and weeping eyes. Her eyes began to itch Feb. 26th; this had been increasing to her great annoyance and was especially severe at night.

EARLY HISTORY.

Has been subject to sore throat since she was a girl, also to severe headaches. For many years has had to clear her throat in the morning, which occasionally made her sick at the stomach.

The treatment gave her immediate relief. She was directed to come the next day, but she did not come until the Saturday following, the 7th. She had but slight return of her symptoms. Was treated again on the 8th, 11th, 14th, 17th and 21st, and not again until Dec. 25th; then once each day until Jan. 1st, following.

She is still in the city, and so far as I know, remains well.

CASE IX.—J. M. C., æt. 29 years, sent to me by Dr. A. B. Barbee, of this city, consulted me Sept 21, 1874, for relief of a severe tickling cough accompanied by symptoms of "hay-fever."

EARLY HISTORY.

Did not think that he took cold when a boy, at least did not know it, but had "running ears" until he was about 17 years old. Has always considered himself as one of the healthiest boys of the family.

One year ago he felt itching symptoms in a slight degree, and was then informed that he was taking "hay-fever." In the early part of this month he had occasion to catch a horse that was in a timothy and clover field, and in doing so became quite warm from running after the animal. About the time he got near enough to the horse " a spell of sneezing would come on," which frightened the animal away from him. He had noticed that his eyelids adhered together in the morning for a few mornings before this attack came on. To the dried, encrusted secretions that adhered to the eyelashes he attributed the intense itching that he had experienced. He did not sneeze more than five or six times, " but the first sneeze closed the nostrils completely." In fifteen or twenty minutes he could breath as freely as usual through the nostrils, and would continue to do so until the next sneezing spell.

He was treated once each day for six days. The first application of vaseline gave him immediate relief. Treated him about three times each week for seven weeks, then twice a week until Nov. 21st.

The next spring he went to Minnesota and remained

free of the complaint up to two years ago, the last time I heard from him.

CASE X.—Miss M. M. N., New Harmony, Mo., æt. 20 years. Sent by Dr. A. Ashford, consulted me June 17th, 1875 for relief from a severe attack of pruritic catarrh (hay-fever).

EARLY HISTORY.

Up to three years ago was very liable to take colds during cold, damp weather. Has suffered for many years with "very severe, blinding headaches," so much so that she could not continue her studies at school. The headache had such an injurious effect on her eyes that she could not read without the aid of glasses. Has had "quinsy sore throat" almost every winter during the last ten years excepting last winter, which was passed without an attack.

For one or two years past, except during the last two weeks before she came to me, she had been unusually free from headaches, colds in the head, sore throat and dyspepsia. This attack commenced on May 31st with short breathing which was occasioned by a tickling cough. At the same time she had weak eyes, which soon began to itch so severely that she occupied her time, for half an hour after going to bed, in rubbing them. The severest sneezing fits were usually after she had been in bed for a few minutes, or until the bed clothes got warm. She wet handkerchiefs by the dozen.

Gave her treatments on June 17th and 19th. These applications gave her so much relief that she thought she could miss one day's treatment, which she did. Was treated about three times a week until the 5th of July, at which time she had a slight "sneezing spell;" she was then treated once a day for ten days, and then twice a week for three weeks. She went home on the 7th of Aug. completely relieved of every symptom of the complaint.

The next year, on Aug. 4th, 1876, she returned for treatment, but had experienced no symptoms of an attack. She had learned to treat herself by means of the spray producers. This had a beneficial effect on her head and throat. During this visit she received fourteen treatments between the 4th of Aug. and the 16th of Sept., at which time she returned home.

As a usual thing patients who attempt to treat themselves make a complete failure of it, but this young lady was quite an adept in handling the instruments. I discourage self-treatment, because of the inability to use the instruments properly, and the lack of judgment as to the quantity of the remedies to be applied.

I have not heard from the patient since.

CASE XI.—Mr. L. M. R., æt. 47 years, a merchant of this city, consulted me Sept. 30th, 1875, for treatment of a pronounced and long standing case of " hay-fever." Every year since 1863 he had to go East, North or West for relief. Had tried almost everything, but found no relief; had no faith in anything except high altitudes.

EARLY HISTORY.

Had been liable to take cold all his life. Never did take good care of himself; is not able to do so now. (It was evident that when he did not have the attack, he did not have the *least thought* of the consequences of his *numerous indiscretions*.) The itching of his eyes almost always commenced his trouble. The dust of his store was his great dread. He said " When I start to sneeze I believe that I would sneeze my head off, if I did not cover it with a silk handkerchief and my soft felt hat. I have tried to see how long I would sneeze without my handkerchief, but I did not have the courage to stand it long enough to see if it would stop while my head was uncovered."

When he first visited me his eyes were very red and his nostrils completely closed. The treatment gave but a slight relief. He ought to have been treated the next day, but I did not " lay the law down to him soon enough." Oct. 2nd the treatment had a very beneficial effect. But it was evident that he was not taking care of himself, except when he " had to." Was treated the 3d and 4th, then missed a day and was treated the 6th and 7th.

The treatment on the 3d and 4th gave him entire exemption, so that he thought he was " not going to have a very bad spell, anyhow ;" but his sneezing returned on the 6th, on which day he was treated. He did not feel at all assured by the treatment on the 8th, so took the cars that night for Denver, Colo., to which place he has resorted every fall since that time.

CASE XII.—Clara T., æt. 8 years. Sent by the late Dr. Frank Porter, of this city, consulted me Sept. 29, 1875. Was first attacked with sneezing on Aug. 22nd, of that year, while she was gathering flowers. At this time she got her face poisoned by " poison ivy," which laid her up in bed for nearly two weeks. After she recovered from this inflammation, the sneezing would instantly commence as soon as she went into the sun or looked up into the sky on a bright day.

PREVIOUS HISTORY.

Always had been a small, nervous child; tonsils enlarged since infancy. Had ear disease and rupture of the membranæ tympani of both ears when about 4 years old; has had no trouble with them since. Slept constantly with her mouth open and made a very loud, choking noise while breathing. For two or three years she had to lay on high pillows during cold weather, to enable her to breathe without disturbing the remainder of the family.

Treated her on the 29th and 30th, Oct. 1st, 2nd and 4th. These treatments relieved her so much that her mother did not bring her back until Nov. 6th, then again on the 15th and 17th. She visited me again for treatment of her enlarged tonsils, March 17th, 1876, and has since remained free from pruritic catarrh.

CASE XIII.—Miss Emma C., of Trenton, Mo., æt. 26 years, consulted me July 30th, 1876, for relief of " rose-fever." She was attacked with this complaint three years ago. The first year the attacks were not very frequent nor severe, but the disease increased each year since. The attack commenced this year on the 4th of July, while enjoying herself at a picnic in the woods. It was so severe that she held her head in her lap for nearly one hour before she could endure the light, her eyes being much more affected than her nasal passages, that is they were far more painful. She was enabled to go home after tying three thick veils over her face and around her head. After she arrived at her home, she had a severe chill and a high fever during the first part of the night.

EARLY HISTORY.

Has had chronic catarrh for many years, and with it a cough every winter.

The first treatment was given with too great an air pressure on the spray producers, and for this reason did not give the relief that I anticipated. Treated her again the next day with the best effect. These treatments were continued once daily for ten days, then three times a week for three weeks, at which time she was unexpectedly called home.

April 4th, 1878, I treated her again for three weeks, three times each week.

I have no knowledge of her condition since then, but have every reason to believe that she has remained well.

CASE XIV.—Mr. James L., a merchant in this city, æt. about 38 years, consulted me on June 6th, 1876, on account of a severe cold in the head. He had been a victim of hay-fever for about four years. Each year his complaint commenced about the 20th of August.

Examination by the pharyngeal mirror revealed nothing unusual except chronic inflammation.

I gave him ten or twelve applications with the spray producers No's 3, 4, 5 and 2. The last application started him to sneezing, which he feared would commence an attack of hay-fever, but it did not. He was relieved of the cold in the head, but received no further applications, preferring to make a visit to the West for relief and relaxation from business.

He returned to my office June 4th, 1877, to be treated for a severe cold affecting his throat as well as his eyes and nose. His trip to Denver, Colo., was productive of much benefit to his health.

EARLY HISTORY.

At first thought he had not been subject to frequent colds while a boy, but upon conversation with his father recollected that he had had scarlet fever very severely when 7 years old, which left him very weak for several years, especially during the winter months. When 20 years old, the late Dr. Pope took a large tumor from his nose; he had forgotten which side ; at that time his mother told him he had a bad breath.

Treated him daily from 4th to the 10th. Then three times a week for three weeks which relieved him.

Recommended him to return during the early part of August, for preventive treatment for his "hay-fever." He did so, and was treated Aug. 9th, 10th, 11th, 12th, 13th, 14th, 16th 18th, and daily until the 23d, at which time he had his attack of hay-fever. On the evening of this day he started for Denver, Colo. As he felt entirely well by the 20th of Sept. he returned.

He visited me again on July 29th, 1878, for preventive treatment. He was treated three times a week for three weeks; commencing on the 18th of Aug. he was treated daily for two weeks. After Sept. he was treated every other day until the 15th, then discontinued, entirely well.

He was again treated a few times during July, 1879 and 1880. Since that time he has remained well as far as I know. As he lives in the city I am sure that he would have returned if he had not remained in good health, as I now have one of his relatives under treatment.

CASE XV.—Mrs. G., æt. 52 years, a German from Quincy, Ill., consulted me on June 26th, 1877, for excessive fits of sneezing. She would sometimes sneeze as many as eighteen or twenty times before stopping, but usually not more than ten or fifteen times. These attacks would come on every ten or fifteen minutes or half hour. As she was quite a heavy woman, these sneezing spells wearied her very much. These attacks commenced five weeks previous to her visit to me, and were constantly increasing.

Examination showed excessive redness of the mucous membrane, and it was much swollen, both nostrils being closed.

Vasoline and three drops of the pinus comp. was sprayed by the No's 4 and 5, vasoline alone by the

No. 2. Three applications were made daily for about two weeks, then every other day for three weeks. A laxative, diuretic and tonic were prescribed.

At the end of the first application her symptoms were *very much* ameliorated, so much so that she had *no more* severe sneezing fits. In two weeks all sneezing ceased, and every symptom disappeared after three more weeks of treatment.

On May 5, 1881, she returned for treatment. After her last treatment, four years ago, she remained entirely well until thefollowingMarch, at which time she took several severe colds, which brought on a shortness of breath, resembling asthma. She now lives near Decatur, Ill., where she thinks she has taken more cold on account of the flatness of the country.

EARLY HISTORY.

Up to the age of 22 years, the time that she was married, she was always sickly. Had sore throat almost every winter, and a bad cough. Had headache until she was about 40 years old. Always had trouble in clearing her throat in the morning, and was sometimes quite sick at the stomach after and while coughing.

I treated ner for chronic nasal catarrh. The treatment lasted until July. The first three days, once daily, then three times a week until June 11th, then twice a week until July 5th.

She has remained well since that time, but has received five or six treatments for her chronic nasal catarrh during Oct. 1883, and once in April, 1884.

These histories of my patients prove what I proposed to do, namely that pruritus rhinitis catarrhus is one of the sequences of chronic nasal catarrh.

CHAPTER V.

LOCAL SYMPTOMS; SUBJECTIVE AND OBJECTIVE.

It is impossible to give the local symptoms, so that they may be seen in every case that may come to the readers observation, for the reason that all symptoms vary according to the age of the complaint and the temperment of the sufferer, but enough can be given to pretty fully portray the peculiarities of the ailment.

THE SKIN.

The skin of the nose and face is frequently the first to be affected by an itching sensation. Sometimes it is a little hightened in color, even before it is rubbed and appears as though a rash was about to break out. Then this sensation extends to the scalp, on the back of the neck, between the shoulders and under the arms. In extreme cases the integument of the whole body suffers to some extent.

After the complaint has lasted about one week, and the skin has been vigorously rubbed in the attempt to relieve it of the itching, an eruption is frequently observed, resembling prurigo. Sometimes the angles of the eyes, especially the inner, become quite inflamed, which the ever-present itching induces the victims to aggravate by more rubbing, until small crusts form on the irritated spots. Slight ulceration appears at the alæ of the nostrils, causing considerable suffering when the itching compels the victims to severely rub the parts for relief. The same kind of breaking out, or herpetic appearance is observed around the mouth.

Some cases suffer from extreme itching on the ankles and wrists, and when rubbed, becomes swollen and sore; then pustules appear, which when ruptured do not quickly heal.

Most patients perspire easily and freely, then the skin becomes excessively sensitive to even slight draughts of air, and become cold and clammy.

A peculiarity of the eruptions is its sudden appearance and disappearance, lasting frequently but a few minutes or hours. When such is the case, the skin is very easily chafed, especially around the neck where the band of the under-vest rubs the parts.

Dr. Wyman mentions a case who "had redness of the skin of the color of a boiled lobster, compelling him to keep his bed five days."

THE EYES.

The eyes come next in the succesion of being the most early and most frequently affected, the itching—the characteristic of the complaint—usually commences at the inner corners. If the left nostril has been the one more affected with the chronic catarrhal inflammation, then the left eye is the first and more severely affected with the itching. The irritation always reddens the conjunctiva, then the whole eye is suffused in tears, the lids become swollen and in the morning they are agglutinated to each other by the Meibomian secretions. On awaking in the morning this instantly gives rise to an attack of itching of the eye-lids, which instantly extends to the nostrils. So "unanimously," as one of my patients expressed it, does this take place, that he was unable to say which part was first affected. This condition of things lasts but a few seconds when the nostrils are completely closed, apparently on account of the tears flowing down the lachrymal canals.

The tears have a positively irritating effect on the cheeks as they flow from the eyes. When the eyes are in this condition, a bright sunlight is so very aggravating that the victims instantly endeavor to shut out the light by placing both of his hands over his face. A dark, cool room is the only place in which he can quickly recover from his attack.

A peculiarity is, that after the attack, the congestion of the blood vessels as suddenly disappears as the attack appeared, leaving no visable trace behind, although in some cases sties are apparently the result of the excessive hyperæmia of the lids.

THE NASAL CAVITIES.

The nose is sometimes the location from which the pruritic symptoms originate. This may be started by a slight push in any direction but especially if given sidewise. I had one patient whose principal agony came from minute boils that formed but did not entirely heal up until the season was past. In some patients the muscles connected with the nose were in almost continual sposmodic contraction a kind of coreaic condition just previous to an attack of sneezing. The nasal passages, according to Beard and Wyman are the parts that most frequently suffer first and most severely. The sneezing is occasioned by the itching. The first wink of the eyes sends the irritating tears down the lachrymal canals which instantly starts the itching, this is followed by sneezing and a largely increased flow of nasal mucous that completely occludes the nasal passages. If the victim blows his nose, as he feels inclined to do, this will aggravate the matter, by causing a full, wedged sensation in the cavities.

It is remarkable that the excessive congestion of the mucous membrane does not more frequently lead to nose-bleed. Dr. Wyman mentions a case who had nasal hæmorrhage, I have not seen one.

As soon as the paroxysm is passed, the passages slowly open so that respiration can be again carried on through them. The nostril that was usually obstructed during the chronic catarrhal stage will be the occluded one during the paroxysms.

As the paroxysms are most severe and most frequent in the mornings, the nasal obstruction will occur at this time of the day also.

The quantity of the nasal discharge, in one morning varies from wetting five or six handkerchiefs to twenty. In the older cases, the secretion is of a watery nature, except at the close of the season, when it is somewhat "sticky" but with those who have had but " two or three seasons of it, " the secretion is always "sticky," and toward the close of the season, the purulent character is quite marked.

In a few cases a spirt of violent exercise, to the extent of producing a gentle perspiration has an opening effect on the nasal passages, and a quieting effect on that days attacks.

In every patient the mucous membrane was observed to be in an excessively hyperæmic condition, and of a dark, purplish-red color. The blood vessels usually plainly visible during the chronic catarrhal stage were not in sight.

The sense of smell is always obtunded, and odors, that before gave pleasure while not causing the least irritation, have usually a disagreeable effect, still they were unrecognizable.

THE PHARYNGO-NASAL CAVITY.

The pharyngo-nasal cavity is always less severely affected than the nasal cavity, but an itching sensation is felt here also. The only means of relieving this part is by coughing, retching and vomiting. All of my patients had the coughing and retching, and most of them had the vomiting.

The mucous membrane while not so deep a red color as the superior turbinated processes, was quite dark red, and in some patients the membrane had an œdematous appearance.

The subjective symptoms due to inflammation in this locality are almost uniformly felt in the throat and for this reason patients try to relieve themselves by coughing.

VELUM AND UVULA.

The soft palate and the uvula are very frequently the seat of an itching sensation. In severe cases, at the end of the season, the velum is frequently in a paretic condition, so much so as to allow fluids to pass up into the pharyngo-nasal cavity and nostrils. In a few cases the uvula is slightly œdematous; in one patient it was so dropsical that it almost filled the whole space between the enlarged tonsils. In this patient the sense of suffocation on assuming the horizontal position, was so great that he slept in an arm chair all night. In some the uvula is so much elongated that it acts as a foreign body in maintaining the cough.

EUSTACHIAN TUBES AND MIDDLE EARS.

The itching sensation sometimes extends up the Eustachian tubes to the middle ears. As soon as these cavities are reached a fine sticking sensation is experienced in the root of the tongue, showing that the chorda tympani nerve is affected. In about a fourth of my patients their hearing was manifestly decreased.

FAUCES AND LARYNX.

On account of the excessive effort to relieve the throat of the itching sensation by coughing, the whole surface is much congested and in an excessively sensitive condition, so much so that it requires some dexterity to make an examination and to apply the spray producers.

A paretic condition of the faucial muscles is sometimes observed, and with this the parts lose their proper sensation, that is become anæsthetic, so that it is quite a labor to swallow food.

THE TONSILS.

The tonsils are not often swollen, but are frequently quite painful, and are particularly so when swallowing. This pain is sometimes felt up in the ears, or if one tonsil is alone affected the corresponding ear is the one in which the pain is felt, and the hearing in this ear is always defective.

When both tonsils are swollen and painful, and the nostrils are closed, eating and drinking is a somewhat dangerous operation, on account of the liability of the food being either driven up into the pharyngo-nasal cavity or allowed to partly pass into the larynx; in the latter event give rise to severe spasmodic coughing and threatened asphyxia

If the nostrils are occluded, so that respiration is carried on through the mouth, the lips, gums, tongue, soft palate and throat all become dry and parched, and all seem as though it was impossible to move or use them, but as soon as a little water is taken into the mouth, and made to bathe all the parts their faculties return.

The secretion from the throat is quite tough if it

is not profuse, and the effect to get rid of it frequently maintains the throat in a sore condition. I have
had but one patient who had severe itching in the
roof of the mouth all the others had this sensation in
this locality but slightly.

THE TRACHEA, BRONCHIAL TUBES, AND LUNGS.

The itching extends from the fauces to the larynx,
and thence to the trachea and lower air passages.
This sensation is the sole cause of the spasmodic action of the lower air passages, or in other words,
the asthmatic symptoms.

The cough does not commence until the parts are
very much irritated by the endeavors of the victim to
relieve himself of the itching. For this reason the
cough is observed in the second and third week of
the pruritic season. The itching is sometimes felt in
the trachea or at least the victim asserts that it is deep
in the chest, where one would locate the wind-pipe.

If the sufferer is awakened by the itching sensation
in his face, eyes or nose, before he gets through at-
tending to these parts with his hands, his tongue is
called upon to relieve the same sensation in the roof
of the mouth, and a rasping cough is raised for the
purpose of relieving the throat, and instantly on this
attempt being made the same sensation is felt in the
larynx, trachea and even in the bronchial tubes.

DECEPTIVE SENSATIONS.

The sensation experienced in the throat, is occasioned by the itching in the pharyngo-nasal cavity.

This is easily proven by the application of a soothing remedy used by means of the spray producer that throws a vertical stream. If this is the case, then it is evident that coughing or clearing the throat will not relieve the irritation located up behind the soft palate, at least five inches above the vocal cords the locality of the cough, and it is also as evident that the less that the patient coughs, the less irritation to the vocal cords, the larynx and the throat, not to mention the effect of a fruitless cough on the air passages in the lungs.

Some patients are so wearied by their efforts at coughing that they can hardly stand; the cough is especially fatiguing if the expectoration is scanty, in this case the efforts at relieveing the throat of the itching sensation is brought about by retching and frequently by vomiting.

THE VOICE.

The voice is soon affected, so that hoarseness is a constant symptom after two or three weeks coughing. The color of the vocal cords is the same as that of the surrounding mucous membrane, instead of being a pearly white resembling the sclerotic coat of the eye.

ASTHMATICS

Toward the latter part of the "pruritic period" the symptoms seem to be less severe in the eyes, face, nasal passages than ever and a slight cough is sufficient to bring on short breathing or asthmatic symptoms. I am satisfied that if the patient could be relieved of the irritation that produces the irresistible desire to cough, the asthma would not follow and those who had but slight or no cough were as free of asthma, while those who early commenced to cough frequently and severely, were afflicted with asthma to the severest degree, in other words, the milder the pruritic symptoms the milder the asthma.

A dinner, made hearty by the use of stimulants, is apt to induce short breathing, but it is not a genuine attack like the one that comes on immediately after the first coughing spell on retiring for the night, these attacks cause the victim to grasp the mantel piece for support.

As the pectoral and intercostal muscles are severely exercised in coughing, this may give rise to a pain in the chest, which may fill the patient with fear least his lungs are becoming seriously involved, but even a slight examination will readily prove that they are not seriously affected, although mucous râles will be easily heard, but this will pass away in a few hours, may be to again reappear after the next paroxysm. The attacks of asthma that follow retching without vomiting always last longer than when there is vomiting. Why? Because the act of vomiting clears out the pharyngo-nasal cavity quickly, whereas the retching alone does not do so, showing that irritation in this cavity can have a marked effect on the lungs, as well as on the larynx.

THE HEART.

Palpitation of the heart is a frequent sequence of of this complaint; so is an intermittent pulse. Most patients complain of a soreness in the region of the heart after they have recovered from their asthmatic attacks. The pulse is not more frequent than would be expected after the bodily exertion of the paroxysms. Many of these patients live under the impression that they have heart disease, but this organ is not affected except in sufferers who have had chronic nasal catarrh for thirty-five or forty years.

CHAPTER VI.

Constitutional Symptoms.

A statement of the constitutional symptoms must, of course, include much that has been said concerning the local manifestations.

A large proportion of these patients are so unobservant of their conditon, that it is with difficulty that premonitory symptoms can be found to exist. A few patients have stated, after being questioned several times on the subject, that they felt as though they were weaker or more nervous; that their appetite was not as good as usual; that their urine was a little more highly colored; that they did not sleep as soundly, and that they felt peevish and cross. Most of these initiatory symptoms were entirely ignored by at least two thirds of my patients. Those who did have any or all of these initial symptoms, state that they probably commenced a week or a little more before their anticipated attack, but were the strongest during the three or four days before the attack.

Some of these patients thought that the mental anxiety concerning the attack, had something to do bringing on these symptoms, together with their loss of sleep and appetite, etc.

In a majoirity of the patients the system was not disturbed until they had suffered for nearly a week. Then they experienced chilliness followed by flashes of heat; their hands and feet burned so severely that they felt compelled to bathe them in cool water, not cold water, as the latter always caused pain. Some preferred tepid water.

While they felt a slight burning heat over the whole of the body, yet when a cool wind struck them especially on the back, they began shivering instantly, this was frequently followed by an attack of itching and sneezing. Several patients were affected with night sweats.

Palpitation of the heart was a very common complaint, as well as a soreness or uneasy sensation in the left side of the chest, especially after sleeping on that side. With many the pulse was slightly accellerated and intermittent.

All patients complain of a want of appetite, not only this, but they lose the faculty of taste to some degree. Warm drinks are the most pleasant, and hot soups keep their place throughout the entire attack, as being the most nourishing and grateful.

With the diminution of the renal secretion, there is constipation of the bowels. Some patients were afflicted with diarrhœa, but this was nearly always found to be due to some indiscretion in eating. Indigestion and all its usual consequences, was almost always present.

All patients were mentally depressed, were fretful

and easily angered, and much given to fussiness about
their meals. Forgetfulness was one of the sequences,
as well as an impossibility to continue a long time in
any train of thought. Many expressed doubts as to
the perfectness of their sanity, or were fearful that
their mind would give way under the terrible stress.
The failing or wandering of the mind was most frequently experienced at night, on waking out of a sleep
when attacked by a sneezing spell. Two patients
were so much terrified by some unaccountable fear
that they would not sleep in the room without a light.
These terrors would even follow them in sleep and
cause them to moan loud enough to awaken those in
an adjoining room. These mental symptoms were
always the most severe with the asthmatics, and those
who had the full attacks or the so-called autumnal
catarrh.

It is remarkable that some victims will undergo
these attacks and suffer from all of these symptoms and
at the end of the season proclaim themselves in good
health, and, because a few people have claimed in
the public prints and in small works on this subject,
that they are better after suffering from these attacks,
they will "follow suit" and make the same expressions, yet every one of them, if carefully interrogated,
will give evidence of the yearly weakening of the
system, that would not have occurred were it not for
these attacks.

CHAPTER VII.

Course of Pruritic Rhinitis.

Some authors who have written on this subject, employ language, in describing the attacks of the complaint, that plainly indicates that they are not averse to using the marvelous, so much so that their remarks concerning its peculiarities, require that the exclammation point should conclude their sentences. Nearly the whole tenor of all they say, is gauged on this key. Even some of the sufferers themselves seem to enjoy this extravagant mode of expression as seen from the histories they give of their symptoms. Brooklyn's world-famed divine takes the lead in this style in describing his case. He shows his fondness of the graphic in detailing his attacks, and one would be really excusable in think-that he was in rather good humor with himself while giving his account of them. I do not mean to question the correctness of what he says, but I insist that the exhibition of the marvelous is manifest and is misleading.

The point I wish to make is this. If a complain-

ant's symptoms do not come up to this marvelous gage,
or his expressions are not given in this key, they are
without exception, not included in the hay-fever list;
consequently, the true commencement of the complaint is not observed, because the first, the initial
manifestations are so very slight, that even the victim
himself does not recognize the tendency of his symptoms, for this reason there is no opportunity for
flights of rhtoric in describing his feelings.

Not long ago, I had a patient say : "I would believe
that I was affected with hay-fever, if I was not so
well acquainted with the symptoms of this disease.
My uncle has had it for many years.   He has it on
the 12th day of June every year, at the same hour in
the morning and it leaves him in just five weeks,
whereas my symptoms last different lengths of time,
and may come on at almost any time of the first four
months of the year."

I know of a gentleman who said: "I had my first
attack on the 20th day of April and it lasted until
frost, at the same time I was in real good health.   I
had no regular time for my itching and sneezing,
neither of which are severe.   I sometimes have it one
time and sometimes another.   One year ago I had it
in February; this year it commenced in May.   One
time I had it in the last week in December;   this was
three or four years ago.   I had almost forgotten it.
I am sure that it is getting worse every year, and
may be it will turn into real hay-fever if it is not
stopped."

This is a fair history of at least thirty cases that have come under my observation.

The forgetfulness of these patients, which is almost proverbial (at least with those that I have seen) makes it difficult to get an accurate and full history of their condition before their attack, but careful interrogations, made from time to time during their visits for treatment, *will always elict the fact* that they had attacks of itching of the eyes, face, nose, etc, with sneezing during the late winter and early spring months, and that they were thus afflicted for from one to seven years previous to the full formation of the complaint.

This view of the commencement of pruritic catarrh can only be established by facts obtained from patients. This I propose to do by giving the histories of a number of patients who have suffered more or less from this complaint.

CASE XVI.—Mrs. A. E. æt. 47 years consulted me Aug. 25 1883. Complained of having a severe cold in the head and disease of the right antrum of Highmore. Since the middle of July 1883, she suffered from what she thought was a continous cold, accompanied by severe sneezing and itching of the eyes and throat. On her first visit she stated that she did not have these symptom at any previous time, but on further conversation on this subject, on the 5th of Sept. following, she remembered having had these " sneezing spells " during the warm weather of the past five years, and that they had been gradually getting worse, these " sneezing spells " were accompanied by symptoms of asthma, which was more severe this year than at any time previous.

CASE XVII.—Dr. R. L. æt. 44 years consulted me June 13th. 1884. He had had nasal catarrh for 25 years, but not very annoying until 10 years ago, since that time, it had been noticable to others that he took cold more frequently in warm weather, so much so that he was not able to attend to business without much suffering. In the spring, when the "fuzz" was blowing off the trees, his eyes became inflamed and all of his catarrhal symptoms were much aggravated. Towards the close of the season he had symptoms of asthma. He was sure that his "hot weather catarrh" had become more aggravated each year, and with it, his asthmatic symptoms.

CASE XVIII.—Miss. M. Collinisville, Ill. æt. 26 years sent me, on June 3rd. 1884, by Dr. Wesseler of this city. Complains of sore throat accompanied by severe cough and short breathing.

In June 1873, upon sweeping the floor or making feather beds her eyes would commence to itch and she immediatly began to sneeze. These paroxysms lasted for nearly an hour; before this date she had the same sneezing now and then, but never so severe. These symptoms did not increase until June 1883, at this time she took a very severe cold. She was certain that she had sneezed every month since that time, especially since last Christmas. At present she sneezes only when she is in a draught. Not otherwise except when exposed to dust from a carpet.

Just before she sneezes she experiences itching over the eyebrows. The symptoms of sneezing are occasioned by emanations from a rose or other flowers.

CASE XIX—H. W. æt 13, years. Consulted me July 3rd. 1883 complained of sneezing, swollen eyes, sore

throat, general debility, nose-bled every day, sometimes several times a day; takes cold easily, especially
during warm weather. When he was an infant about
3 or 4 weeks old he took a severe cold.

His mother thinks this was the foundation of his
catarrhal trouble. In three or four months after this,
took another severe cold and had inflammation of the
pleura. He did not take cold like most childern but
had a watery discharge from his nostrils like an adult.
Fresh hay did not have any irritating effect, but old
dusty hay always brought on paroxysm of sneezing. His mother does not remember his sneezing con-
tinuously as he does at present, but knows that he has
been rather sensative to all kinds of dust since he was
three or four year old. Had not had attacks of
asthma until the fall of 1882. All of his symptoms
remained the same until 1880. Since then they have
been increasing rapidly.

CASE XX.—Mr. A. A. æt 17 years complained of
sneezing and itching of the eyes. When he was 8
years old he caught a severe cold in the cars, that affected him for nearly a year. On July 18th 1875,
while in Concord, N. H., he went into a hay field and
was attacked with sneezing, which was so severe as to
compel him to return home. The paroxysms did
not fully discontinue until evening, at this time a
feather dipped in quiuine was put up his nose, but
this made him sneeze more severly, and had the
effect of maintaining the attack for nearly the whole
of the night. This application of quinine was kept
up for some time, with the effect, as his mother now
maintains, of producing a chronic inflammation of the
nasal passages. On the discontinuance of the quinine his paroxysms ceased. At first his physician
thought that he had hay fever, but as he remained

well and was able to pass through hay fields and play with hay it was concluded that he did not have the complaint. The next attack of sneezing occurred in the later part of May 1876. On its reappearance some of his friends thought he had hay-fever, but after consultation with a physician it was concluded that it was only a cold in the head, the next year he again had the paroxysms in May, at this time he was sent to Iowa where in a few days all symptoms disappeared. In 1878 he had but few paroxysms of sneezing. In 1879 the complaint developed itself in full force, for which he was treated by three physicians. In 1880, his attack was postponed until the 15th of June, and for three weeks it was very severe, on the 12th of August in the same year it again commenced and continued for about one week, towards the end of September he had a third attack which lasted him for two weeks. In 1881 he did not experience any sensation until the end of September. In 1882 had it slightly in August, and a severe attack of it the first week of October.

CASE XXI.—Mr. W. H. M. æt 42 years, consulted-me May, 12 1885. "For three years I have been sub ject to taking cold far more frequent in warm spring weather than during winter. My colds have been so severe that I have lost my voice. My eyes have been weak and watery at such times and I think they itched last spring and may be a little the year before, but at present I have to rub them to relieve them of itching which is now quite severe. I have sneezed a good deal for several years, but had not thought of having hay-fever."

The above histories plainly establish the fact that pruritic catarrh manifests itself by slight symptoms at first, and that these gradually increase in severity until it takes full possession of its victim, or in other words, is so violent in its demonstrations that it forces recognition from every one.

The course of the complaint after it has fully manifested itself, is of interest to the physician. From it he can determine whether or not the complaint is decreasing or increasing under his treatment.

To describe the course of this complaint, I will be compelled to take the stages, consecutively, as they occur.

STAGE OF NON-RECOGNITION.

This is the most important stage; it commences at any time from the last week in December to the first of July; but the period for the most frequent attacks is in May and June. The victim has been subject to colds in the head for years; he has indulged in reading at night or has smoked at night, then his eyes itch him a little, which he may blame to the smoke of his cigar; he has sneezed a little. Even if he is not the age to use glasses, he will be apt to question himself concerning the failure of his sight, and will seriously think of doing something for this disability. He will notice that they are sticky in the morning, and that they are apt to water if he suddenly goes into the light, also that this may be accompanied by a few sneezes.

If these symptoms are marked, the usual symptoms that he experienced of his chronic nasal catarrh will be

proportionably lessened, showing a marked metasta-
sis of the former complaint to the new one. Indeed,
this characteristic is observable in all stages of chronic
catarrhal inflammation of the nasal cavities and all
of its sequelæ, pruritic catarrh not excepted. I had
one patient who imprudently ate some canned apples
in June, at a time he was suffering from his attacks,
which gave rise to a bowel complaint that resembled
cholera morbus. While ailing with this disease, he
was entirely free from his slight symptoms of pruritic
catarrh, and from his severe headaches, the result of
chronic nasal catarrh.

These irregular attacks of itching and sneezing may
last for a few months or may be for several years, but
when they assume a severer type, they then take on
more regular dates of commencement, and discontin-
ance, and the complaint is then given the name of
rose-cold or June-cold or July-cold, according to the
season of the year in which these regular attacks
occur.

This brings us to the recognized

PRURITIC CATARRH, OR THE EARLY FORM KNOWN AS ROSE-COLD OR JUNE-COLD.

The facts herewith given will show plainly that the
earlier the attacks of pruritic catarrh, the younger the
complaint, and *vice versâ.*

While these views are not acquiesed in by any author that I have seen, yet I will quote from them passages and histories of cases, that will prove that I am
right. The two symptoms, the cough—which comes

from pharyngo-nasal irritation—and the asthma, that are taken to show the degree of the severity of the complaint, increases as the age of the complaint increases. This is shown by Dr. Beard, on page 111.

Of 17 May cases, 4 had neither cough nor asthma, 7 had both.

Of 13 July cases, 3 had neither cough nor asthma, 7 had both.

Of 55 August cases, 9 had neither cough nor asthma 66 had one or had both, demonstrating that the later in the season the attacks occur, the more severe the complications. Further on he says: " Others, who during the first years have the early (May) or middle (July) form, subsequently have the late form (August).

In corroboration of this, I will quote from Dr. Wyman. He gives on page 148, the histories of cases exactly similar to quite a number that I have under observation viz :

"Dr. A. W. W. of Chicago, Ill.—Has suffered since his eighteenth year, though for ten years it took the form of " rose cold" or June catarrh. 'Finally six years ago the June visitation was broken up by Jonas Whitcomb's remedy, and I was congratulating myself on a cure when August came and brought with it the 'big brother.' Since that time I have no further trouble in June—save it all for August.'

On page 149 Dr Wyman says : "The June cold is less severe and of shorter duration ; the eyes are less severely and less constantly affected; the cough is much less constant, and not spasmodic to the degree of producing retching and vomiting ; asthma is less frequent at the close, but when it exists is sometimes more severe. * * * * * * *

"Those who have June cold are seldom subjects of Autumnal catarrh. When June cold has existed it has generally ceased on the appearance of the latter disease."

On this page he gives the history of some cases illustrating what I have quoted: "Mrs. H. at the age of 18 first noticed that she was affected by the aroma of roses. The following year, while picking roses in the morning, had itching of the eyes, which became so intolerable by afternoon that she asked medical advice. After this she could not be in a room with any flowers without affection of the eyes and catarrhal symptoms. This state of things continued about ten years, when she began to have her regular Autumnal catarrh, and the sensitiveness to flowers very materially decreased, but has not entirely disappeared."

On page 150, is the following cases:

"*Case 40.*—The yearly attack formerly commenced in June; now it commences between August 20th and 27th, and terminates September 10th to 20th."

"*Case 65.*—Mrs. M.—At 16 had catarrh commencing in June and ending about July 4th, or during haying time. This occured annually for 17 years. Five years ago, after some irregularity in its period of termination, it ceased altogether, and a catarrh appeared about August 1st, when near Fall River, Mass. The three subsequent years she was in Origan, Illinois, when it appeared August 17th, and this year [1866] while in Charlestown, Mass., August 24th."

On page 151 is the following case:

"Rev. J. H. W., who had been a subject of June cold from early infancy, writes: 'But it has changed. It always had begun in June and continued until the middle of July; but about ten years ago it began to reappear in Autumn. Now it has almost transfered itself from June to September. *E. g.*, this year [1872] I have had two bad days, one in June and one in July. Last year it was about the same ; but with September came three terrible weeks, part of which I had to 'give up' and take my bed, for the first time in ten years of preaching I lost a Sunday's duty from this cause."

Many other cases of the same character could be quoted from these and other authors. I have the histories of 23 patients that bear the same testimony.

This demonstrates the difficulty in giving the course of this complaint, as it differs according to the age, temperment and exposures of the victims.

Once that the complaint has sufficiently marked symptoms, the patient will complain of an itching of the eyes, but this always occurs after an exposure that has resulted in a cold being taken.

If the itching lasts for a few minutes, the same sensation is experienced in the nose, then the sneezing comes on, first only a few of them, not enough to call his or her friends attention. The next day this same course is experienced by the victim and is observed by his friends, then (in the majority of instances) they tell him that he has the hay-fever.

At first his attacks are not severe, and after they have passed away, he *invaribly* thinks that "it may not be the hay-fever, after all." I have not seen the victim of pruritic catarrh that did not say this or use words to this effect; not only this, if they pass one day,

they as invariably forget to take any precautions to
prevent a recurrence.

After they have had an attack for about a week or
two, they may begin to have a cough, but this is far
from being constant, except when the complaint has
grown to be a few years of age, then asthma in a
mild form may affect the patient.

All their previous symptoms of chronic nasal catarrh
will disappear as these new symptoms appear, yet the
patient will not mark the absence of his old symptoms unless reminded of it by some friend or some
unusual circumstance.

At this age of the complaint, there is no premoni-
tary stage, at least very few patients can remember
any thing that would resemble that experienced by
older victims.

The duration may be a few days, or one or two
weeks, according to the care the patient takes of
himself, and to the exposures that he must encounter;
about ten days is the average duration. If the first
on-set passes off in three or four days, his tormentor
may return with redoubled force. The date of dis-
appearrnce is always uncertain in young cases, and it
may be sudden or it may be gradual.

THE ATTACKS THAT OCCUR IN JULY.

I do not mean, by thus dividing the description of
the attacks of this complaint, that they are different
diseases but a difference only in   severity. A good
illustration of what I mean is to say that the intermittent, remittent and continued fevers are not three

separate diseases, but three grades of severity of the
same disease. On this point I agree with Dr. Beard
he says on page 110, " The unity of the different forms
of hay--fever, occuring early in the summer, in mid-
summer, or late in the fall, is proved by the following facts :

" The symptoms in all three forms are the same in
kind, differing, if at all, in degree only. The distinc-
tive symptoms—the sneezing, itching, discharge from
the nose and eyes, swelling and obstruction, cough
and asthma, with the febrile state, nervousness, lan-
guor, debility, and depression—are experienced in
the early and middle as well as in the latter forms. "

As seen from the tables showing the dates of attack
[see index], the number attacked in July are not
greater than in June, but the severity of the attacks
are always greater.

Some might say, that according to my theory the
number also ought to be greater. It does look that
way at first sight, but this can easily be explained.
The early attacks are made more frequent by the
greater liability to take cold, because the season of
the year tends to cold-taking. Many of these attacks
would not have occurred were it not for some indiscretion that could have been easily avoided, and would
not have happened in warmer weather if the victim had
taken care. Thus the number of sufferers would be
decreased, in proportion as the mildness of the season
made it possible for indiscrete persons to be careless
of hygienic measures, without taking severe colds.
The symptoms of this form, the July attacks, do not

differ in the least except as to severity and duration. More of them have cough and asthma and the attack lasts longer. To repeat them would lead to confusion.

THE ATTACKS THAT OCCUR IN THE AUTUMN.

This is THE SEASON for the attacks of this complaint after it is well formed, but the symptoms of this state also differ only in degree of severity and duration, that is, with the average number of the cases. Some of the mild autumnal grades are very much less severe than some of the severe forms of the July or May forms. But few victims of this form escape the cough and asthma.

A peculiarity of this stage is that the victim sometimes out-lasts the complaint, that is, the attacks, after coming on regularly for a number of years, slowly decrease in severity and then cease altogether. I have seen two cases of this kind. Dr. Beard relates three cases that "finally disappeared entirely."

CHAPTER VIII.

INFLUENCE OF LOCATION.

Most of my patients were best pleased with the effects of the climate of Colorado, several of whom had visited the famous White Mountains. The sunshine in the Colorado region does not have the least irritating effect on the eyes. Several were greatly relieved by a sojourn near Austin, Texas. Several have made their residence near Los Angelos, Cal., because of complete exemption from the attacks. One of these patients, a physician, says that he has never known of an individual having an attack of pruritic catarrh while remaining in that region. Three of my patients preferred the region of Lake Superior, Sault St. Marie, and Mackinaw.

Last year (1884) in August and September, three of my patients spent the whole pruritic season at home, in a darkened room, maintained cool by hanging a piece of ice, ten pounds, each day from the ceiling of their apartment, as hereafter described.

Dr. Beard asked his correspondents the following question. "Are you better in the city or in the country?"

His replies were the following :

Better in the city - - - - 28
Better in the country - - - 12
No difference - - - - - 8

The remainder—162—not answering the question in any way. This is so unsatisfactory that I consider it of little or no value.

His next question was : " Where do you find the quickest and surest relief ?"

His replies were the following :

At sea - - - - - - 8
At the sea-side - - - - - 19
In mountainous regions - - - 35
Some portions of White Mountains 15
Rocky Mountains - - - - 7
In bed in a cool, close, dark room - 8

As before stated, three of my patients expressed the same preference that the last eight did, and one of these had been to Denver, Colo., and one to the Tennessee Mountains.

Another of Dr. Beard's questions is the following : " Have you ever visited elevated regions without benefit ?"

His replies were " 83 No. and 17 Yes."

In his detailed replies, the following places had been visited *without benefit*.

" Five, specified Catskill Mountains ; Lenox, Mass ; White Mountains : Little Mountain O.; Mountains of Eastern Pennsylvania and Mt. Mansfield ; Overlook (3000 feet high) ; Mountainous regions of Connecticut ; etc."

Dr. Beard very correctly says : " The information

here contained is sufficient to show that there is no rigidly defined non-catarrhal line. Elevation is but one factor."

FOR VERY CHRONIC CASES.

Those victims who have had the pruritic catarrh for TEN or more years, would do well, after receiving the treatment for their chronic nasal trouble—as all of them have had this form of catarrh,—to visit and remain at a location that affords them complete exemption from its attacks, because the more frequently they allow the attacks to occur, the longer the attack will will remain and the more surely will it fasten itself on them. I believe that a course of treatment, for several years, followed by a prolonged visit to some region in which they can be free from attacks will ultimately cure the complaint, except in very old patients.

Under these circumstances, where will they go?

I have recommended my patients to go to one safe region one year, and another the next year, and so on; that is, not to visit any one place, any two years in succession, as the patient may become acclimated to the place, and thus loose some of the benefits of the change.

OCEAN TRAVEL.

A voyage on the ocean is the most certain to be healthful to these patients. But in so doing one need not take a passage on a vessel that is transporting a large drove of cattle, sheep, hogs etc., and in this way "tempt the evil one." Those vessels that are freighed with pine lumber, or black walnut lumber, or timber, are the best of all. A voyage in a vessel, leaving a little before the beginning of the season in which the pruritic catarrh commences, bound for some part in South America, say Rio Janeiro is a most pleasant, trip, and one that need not be so very expensive either, a good thing to be kept in mind, by the way.

WESTWARD.

The next best regions are in Colorado or California. No bad reports have come from either of these States. While residing in either of these locations, the patient should not conclude that he can, with impunity, live indifferent to the laws of health. One of my patients, who visited the " Far West " said that he " was fool enough to acts as though the climate took the contract to cure him," consequently he followed his usual inclinations in living carelessly, in utter forgetfulness of all hygienic rules. The consequence was that he did not receive all the benefit that he might have done had he lived differently.

LAKE SUPERIOR.

The regions around Mackinaw and Sault Saint Marie are reported as excellent for patients afflicted with this complaint. The air of this part of the country, like the water of this grand lake, is remarkable for its clearness, but even here patients must not expose themselves to agencies that are known to be irritating.

A TOUR ON THE CONTINENT.

This is most always very beneficial. Many have been permanently strengthened by a three or four

months visit to Europe.   In these days of rapid and
pleasant travel, a trip can be made by a patient who
is very weak, with an almost certanity of immediate
relief as soon as land is out of sight.

A permanent residence in Europe is a sure preventive of further annoyance from this complaint.   Although it is stated that when Americans return to this
country, the pruritic catarrh also makes it return, yet
I am fully of the opinion that were they treated for
the chronic catarrhal inflammation of their nasal passages while there, for a period of three years according
to the age of the patient, they could return to their
usual abode with immunity from the attack of the
pruritic catarrh.

In closing this chapter I would again urge that
greater benefit will be received by the traveling victim visiting one region one season, and another the
next season, always looking for a new place of exemption.

CHAPTER IX.

CAUSES OF THE PAROXYSMS.

DUST.

Dust of various kinds stands at the head of the list of the causes of paroxysms, and the dust of the steam cars is the most aggravating; as this is always accompanied by the sulphourous smoke from the locomotive. The next kind of dust that is to be avoided is that from an old carpet, then that from an old feather bed, then from a moss bed, after this comes the dust from old, mouldy hay and from the street. It must be born in mind that in the formative and early stages, the cause of the paroxysms is not attributed to dust alone, there must always be a tendency and one or more other irritating agencies at work, as sunlight, heat, excesive exertion, sufficient to cause perspiration to contribute to this result; but dust, of the kind named, seems to be the most prominent.

BRIGHT SUNLIGHT.

Any kind of very bright light, but especially strong sunlight stands next in the list. If the victim should lie with his face to an unshuttered eastern window,

and suddenly open his eyes so that the full morning
light will fall upon them, an attack is almost certain.

EXHAUSTION FROM HEAT.

Over heating the system stands next as an irritating
influence, but with this there must also be over-exer-
tion to the extent of exhaustion. As these patients
are easily wearied, even moderate exercise may lead
to exhaustion.

The remainder of the list of irritating agencies,
named as they have proved to be the most noxious,
are as follows:

Hay, old and musty, and fresh.
Sneeze or rag-weed.
Sulphur matches.
Smoke.
Draught of damp air.
Flowers.
Air of a mouldy room.
Cold damp winds.
Tobacco smoke.
Foggy morning.
Night air.
Damp cloths.

Of the mental conditions: manifestation of excessive
ill temper; anxiety, and melancholy, are the most
prominent. Indigestion is a frequent excitant of an
attack.

The sudden, in fact the instant response of the schneiderian to the irritating effect of the most of these
agencies, apparently leaves no period for the incuba-
tion of parasites. Notwithstanding this, I presume
that, some one will soon lay claim to the discovery of

bacilariæ peculiar to or may be a cause of this complaint. The effects of these irritating agencies are so instantaneous, that there is no opportunity for imagination to act on the victim, as the attack is a surprise to every one of them, nor do they know positively, for some time, the cause of their paroxysms.

A close investigation of the effects of these irritating agencies, proves that their number is quite large, instead of being but a few things, and that hay in any form is not the chief cause of the paroxysm, dust of various kinds being far the most noxious. This disposes of the question as to the propriety of calling the complaint, hay-fever.

CHAPTER X.

The characteristics of increasing severity of the successive stages of this disorder, is plainly demonstrated by the TABLE facing this page.

THE FIRST OR FORMATIVE STAGE.

Presuming that the reader has scanned this table, the work of presenting the diagnosis will be materally shortened.

In proportion as prevention is more important than alleviation or cure, so is it important that a diagnosis of this complaint should be made as early possible.

The only disease that might be taken for the formative stage of pruritic rhinitis is a common cold in the head. Both complaints are frequently accompanied by sneezing, but with a cold, there is no itching of the face. If itching ever so slight, should be expierenced and it occurs during warm weather, then the complaint may properly be called the first stage of pruritic catarrh.

A common cold does not attack its victims suddenly and it may occur at any season of the year;

DIFFERENTIAL DIAGNOSIS OF THE VARIOUS STAGES OF PRURITIC RHINITIS.

COLDS IN THE HEAD.	FIRST OR FORMATIVE STAGE.	SECOND STAGE OR MAY AND JUNE FORMS.	THIRD STAGE OR JULY FORM.	FOURTH STAGE OR AUTUMNAL FORM.
No itching of the eyes, nose or throat.	Very slight itching of the eyes, nose and throat, especially after a prolonged cold.	Itching of the eyes, nose and throat always present.	Itching of the eyes, nose and throat one of the most prominent symptoms; the skin is also sometimes affected.	The most marked symptom is the itching of the eyes, nose, Eustachian tubes, ears, pharyngo-nasal cavity, pharynx, larynx, trachea, lungs and the whole body, the latter is frequently affected with an eruption as a sequence.
Sometimes sneezing, but never in particular paroxysms; not more than three sneezes at a time.	Sneezing more severe than with an ordinary cold, frequently five to eight sneezes at a time.	Sneezing quite a prominent symptom, paroxysms occurring four or five times a day. Each sneeze is complete. Patient able to keep the handkerchief to his face.	Sneezing is so frequent that it is quite fatiguing; it is in severe paroxysms. The sneezes are partial or incomplete. Patient sometimes does not keep the handkerchief to the face, sneezes without it.	Sneezing is a formidable symptom; some patients fall from their chair during the paroxysm. Almost no sneeze is a complete one. Patient never pretends to keep a handkerchief to his face; grasps a chair or table and sneezes without anything to his face.
Eyes but slightly, if at all affected; light does not effect them in the least.	Eyes more than usually suffused with tears but not inflamed, light has a slightly disagreeable effect.	Tears flow profusely from the eyes, but the eyelids are only slightly inflamed; light has a disagreeable effect.	Eyes suddenly suffused with tears; eyelids still greater inflamed, and are puffy; light even after the cessation of attack is disagreeable.	Eyes blinded with profuse lachrymation, eye-lids greatly inflamed and quite puffy; light cannot be tolerated at any time during the most of the pruritic season.
Not aggravated by dust of any kind; or the emanations of hay or any kind of flowers.	As far as observed is slightly affected by dust, hay and flowers, but irritation is not observed until the subject is mentioned.	All experience a suffocative sensation from dust, and one-half of them are affected by hay, roses and early flowers.	A little more than one-half of them are affected by hay, very few by roses or early flowers; dust from a carpet is the most irritating.	Very few are affected by hay or roses or noses of them are early flowers, but every one of them are affected badly by dust, especially from an old carpet and from cinders; the latter has an almost instant choking effect.
No regular time for attack.	Time of attack, January to July.	Time of attack in May and June.	Time of attack in July.	Time of attack from August to November.
No asthma or spasmodic cough.	Asthmatic breathing and severe cough, but not spasmodic.	Asthmatic breathing and slight attacks of asthma and severe and sometimes spasmodic cough.	Asthma more common and cough more severe and nearly always spasmodic if it lasts long.	Severe attacks of Asthma especially toward the close of the season, cough is always spasmodic.
Not relieved by a change of residence.	No observations made on locality or residence.	Dr. Wyman says that it is generally relieved by a residence at the sea-coast and in large cities. My small experience agrees with this.	Dr. Beard did not make special inquiry as he mixed all forms in his remarks, but it appears that a northern climate and mountainous regions are beneficial.	Dr. Wyman says that entire relief is found in certain mountainous regions.
No certain time for disappearance.	A dose of quinine will cause it to disappear.	Disappears in four or five weeks after it commences.	Disappears in about six or eight weeks after it commences.	With the great majority it disappears with the first hard frost.

but should it occur more frequently in warm weather, then pruritic catarrh may be suspected, especially if its attacks are more sudden than usual
colds, and if accompanied by redness of the eyes and
a profuse flow of tears.

As stated, a simple cold in the head is more liable
to occur at a season of the year that pruritic catarrh
does not occur, yet it must be remembered, that the
pruritic complaint is always preceded by symptoms
of a common cold, and is usually, nay, almost universally taken for a simple cold for a year or two.

While Dr. Beard, in common with all authors, ig-
nores common catarrh as the originating disease, yet
he has passages in his work that fully confirm my views
in this regard; he says: " In the first attack there is
always a doubt which may not be settled until the following year. During the first attack, the severity and
obstinacy of the symptoms and the season of the year
are the chief causes that excite the suspicion of hay-
fever. Those whose first attacks are in infancy or early
child-hood may not suspect the real nature of their
disorder until they arrive at maturity."

To repeat: if the suspected symptoms be accompanied with itching of the eyes, however slight, and
with sneezing, and the attack be sudden and especially
if all this occurs in warm weather, when colds are not
liable to be taken, then it is altogether likely that
the victim is suffering from the first, the formative
stage of Pruritus Rhinitis Catarrhus.

THE SECOND OR THE MAY AND JUNE FORMS.

At this stage, the disorder has shown itself plainly. It now stands in marked contrast to the symptoms accompanying a cold in the head.

An ordinary cold is far more liable to occur at those seasons of the year in which this ailment does not usually occur; a cold comes on gradually, pruritic catarrh suddenly; the eyes may sometimes be a little reddened in a cold, in this complaint they are almost always quite red.

The pathognomonic symptoms of pruritic catarrh, the itching, is not present with a cold; sneezing is observed in both complaints but far more severe in the former; a cold will not disappear so completely in a few hours as will the symptoms of this ailment. Asthmatic breathing very rarely follows a cold, but it not unfrequently follows pruritic catarrh even in this stage.

Dust does not make a cold in the head worse, at least it does not show marked increase because of it, whereas pruritic catarrh is almost instantly made worse by it.

Usual medical treatment, such as a foot bath, a sweat a dose of quinine, etc., will cure a cold, but with this complaint it has but little ameliorating effect. A cold has no fixed time to disappear, the other disappears in four or five weeks. The pruritic catarrh is frequently relieve by change of residence to certain parts of the country, a cold is not. A cold is aggravated by cold weather the other is frequently improved.

Nearly all of these contrasting symptoms are well defined.

THE THIRD STAGE OR THE JULY FORM.

There is still less liability for mistaking this form of pruritic catarrh for a cold in the head; all of the features of the former complaint stand in marked contrast. The season of the year in which it occurs, being such that colds are not liable to be taken, even if patients are quite careless in observance of the laws of hygiene.

Ordinary asthma might be confounded with it by the uninitiated, but the absence of the itching of the eyes, nose and face would show the mistake, besides, with asthma there is much greater impediment in respiration than the pruritic catarrh, except when asthma   is a sequence. With asthma a cool draught of air from an open window is very refreshing, with the other it would be very aggravating.

The asthma that accompanies pruritic catarrh, is always preceded by the usual itching, sneezing etc. while in ordinary asthma, no such symptoms ever occur. The itching and the sneezing that precedes the asthma of pruritic catarrh are the only symptoms that distinguish it from ordinary asthma, in all other respects they are identical.

THE FOURTH STAGE OR THE AUTUMNAL FORM.

This stage is so peculiarly phenomenal that none but the most obtuse observer would take it for a common cold.

Every person that I have seen who had this form

of this complaint has had attacks of the earlier forms.
These facts would exclude all complaints that resembled it in the least.

Dr. Wyman, because " its existence has been doubted, and still is doubted by many, even by physicians, "
has taken special pains to give the differential diagnosis between this and a common cold and acute bron-
chitis, and has also given the points of difference between it and pneumonia and local inflammation of the
eyes, but it seems to me that the physician who would
mistake the one for the other must be a very poor
observer indeed ; just as likely would an educated medical man confound an intermittent fever for a typh-
oid fever.

The diagnosis of each of the four forms have now
been given, and it seems to me that a comparative
study of the symptoms of all of these grades, from
the formative stage through the final or Autumnal
stage, must convince every one that this is but one
complaint, an ailment that progresses in severity,
starting from a common cold in the head, showing itself but slightly in nasal catarrh, then assuming a little more severe character in the May and June forms,
then still increasing in severity in the July form, and
finally culminating in the severest form, the Autumnal.

THE PROGNOSIS.

This will be goverened to some extent by the stage
or form that afflicts the patient. In in the first or
formative stage, nothing is easier to check, I fully be-

lieve that this can be effected in patients not over fifteen years of age, by hygienic measures alone. All
cases not over thirty five years of age will be cured
in one or at most two seasons of treatment, while
with those over forty years of age, it may take a year
or so longer. Every patient that takes good care of
himself will in time completely recover.

In this stage I would not recommend a surgical
operation in any case, in either old or young patient,
as the scar following the galvano-cautery, or caustic
acids, would be almost certain to be followed by a
recurrence of inspissated masses to be blown out of
the nostrils or hawked out of the throat, as soon as
they attain sufficient bulk to impede respiration.
These accretions will be just the size and shape of
the scar. Besides being a source of very great annoyance it might—not always by any means—produce an
unpleasant odor to the breath through the nostrils.
Persistant employment of the spray producers will
cure them.

My experience in the treatment of the second stage,
or May and June forms, leads me to say that they
also will finally recover, but as the complaint has a
stronger hold on them, a longer time for treatment
will be required, and surgical interference may be
needed to bring about the desired relief. But it must
be kept in mind, as indicated above, that the smaller
the operation, the better the ultimate recovery. The
more confirmed the complaint, the longer will the patient require fall and spring treatments to completely

eradicate the primary or originating disease, namely: chronic catarrhal inflammation of the nasal passages.

The same may be said of the third and fourth forms, that is, they will require more " chronic treatment " as it were, and surgical measures will be almost certain to be required. The later in the year that the victims is attacked the longer the time will be required for its treatment and the greater will be the need for surgical interference.

My experience in the treatment of the Autumnal form, justifies me in saying that the course I have laid down in this work is followed by satisfactory results·

CHAPTER XI.

TREATMENT; MEDICAL AND SURGICAL. *

The treatment of pruritic rhinitis is divided into preventive, alleviative and surgical.

The preventive treatment embraces the hygienic and therapeutic treatment for chronic catarrhal inflammation of the nasal passages. † As catarrhal disease has prepared the patient's nasal mucous membrane so as to render him liable to take the complaint, his ultimate recovery will depend upon his being treated for the primary disease fall and spring, or at such times of the year as he is most liable to take cold.

ALLEVIATIVE TREATMENT

The alleviative treatment is divided into therapeutic and hygienic, and the therapeutic into local and constitutional. The consideration of these divisions will be taken up in their natural course as they would occur in the treatment of a patient on arriving at one's office.

The patient has suffered a few days or weeks tor-

---

* Read before the St. Louis Medical Society, May 17th., 1884, and published in the St. Louis Medical and Surgical Journal, Aug. 1884.

† The reader is referred to the work on the "Hygiene and Treatment of Chronic Nasal Catarrh" by the author, pp. 473.

ment, and his nasal passages and throat are excessively sensitive. In examining his pharyngo-nasal cavity, do not cause him to retch or cough ; while inspecting his nasal passages do not make him sneeze by either thrusting the nasal speculum too far up his nostrils or by pushing his nose upward or sideways. Be very careful to avoid doing *anything* that will cause him to sneeze. If a window or door of the office is open, close it immediately, as even a slight draught of air will induce an itching sensation of the face, eyes and nostrils, which, if it lasts beyond half a minute, may bring on a paroxysm in full force.

LOCAL APPLICATIONS.

If a paroxysm does ensue, and indicates that it is going to last for several minutes, give the patient a little vaseline and direct him to anoint his face, neck head—if his hair is short—and hands ; rubbing the vaseline well into the skin ; have him put a silk handkercheif over his head, and his hat over that; then direct him to pull off his boots and socks, and rub his feet well with vaseline, uncovering one at a time. It is altogether likely that his feet will be found to be damp with prespiration consequently quite clammy and cold. In this condition the vaseline will prove a very valuable remedy.

This anointing and rubbing process, will materially shorten the paroxysms and lessened their severity. It will be well for the patient to repeat this course at such times as he may be attacked with a paroxysm.

If ready to make a local application, give the patient
the tongue depressor, tell him to place it well on his

Fig. 1.

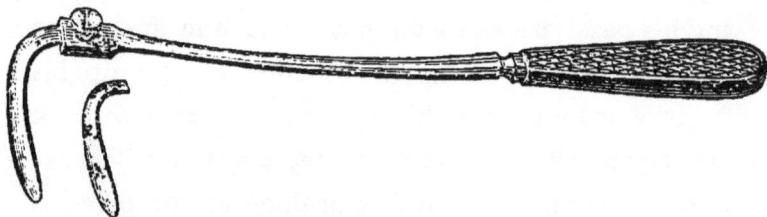

TONGUE-DEPRESSOR.—The patient alone uses this instrument during
examinations, applications and operations. It should be placed well
on the tongue, but not so far back as to cause a retching sensation.

tongue, but not so far back as to cause him to cough
or retch, then, having warmed the spray producer

Fig. 2.

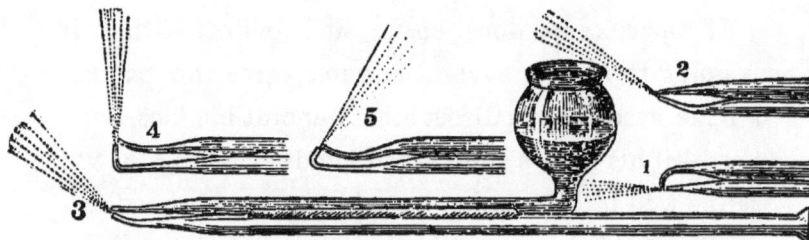

SPRAY-PRODUCERS.—No. 1. throws a stream in a horizontal direction;
applying the medicament to the fauces and tonsils, and by slight inhalation, the larynx and trachea are treated. No. 2, throws a stream at an
angle of about 27° and is used to apply the medicament to the anterior
nares as seen in Fig. 3. No. 3, is not usually employed for treating
pruritic catarrh, but may be used to treat the posterior wall of the pharynx as seen in Fig. 3. No. 4, throws a stream vertically, and treats the
superior portion of the pharyngo-nasal cavity. No. 5, throws a stream
at an angle of 45° and is made to enter the poterior nares, also shown
in Fig. 3.

(No 4), half fill its bowl with plain vaseline and
about 5 grains of the following:

℞. Vaseline........ ....ℨij.
Eucalypti ext. (Sander & Son; Sanhurst
Australia.) min............................. ....x.

Mix while cold.

Fig. 3.

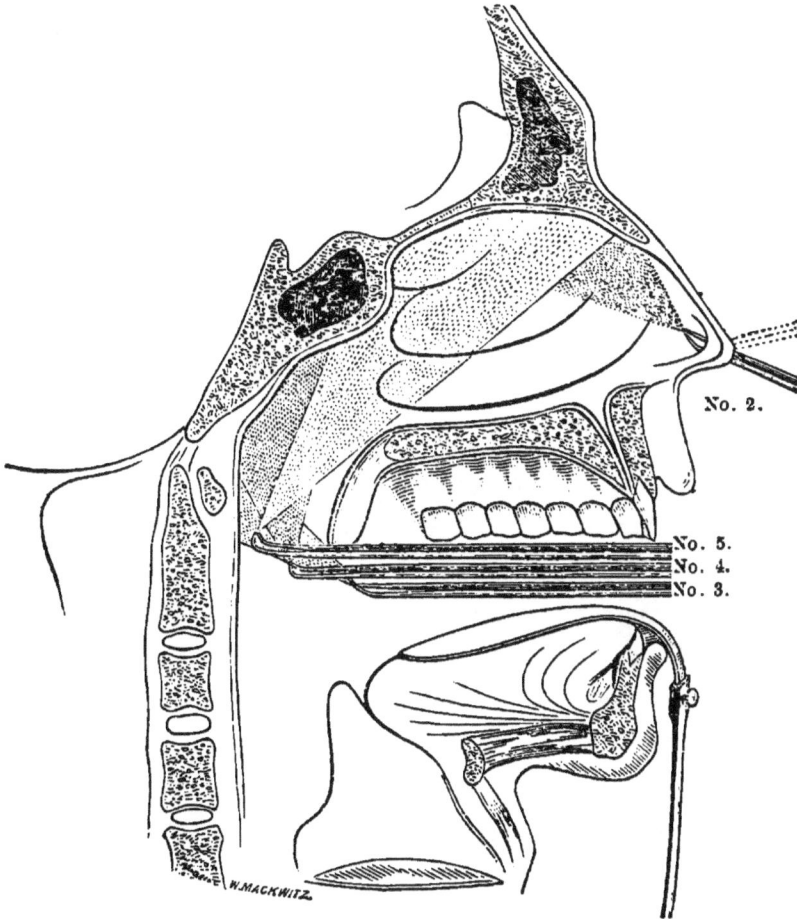

Antero-Posterior Section of the Head, showing the combined direction of Spray Producers Nos. 2, 3, 4 and 5 in the local treatment of the pharyngo-nasal and nasal cavities. No. 2 is introduced into the anterior nares.

The mixture, after it is placed in the spray producer, should be made so hot, that after it is tempered with the cold air from the instrument, the spray will produce a warm, pleasant sensation. Place the point of the spray producer just behind and below the pendent velum and alternately to each side of the uvula; throw the spray up behind the soft palate, gently at first (never using a pressure exceeding 7 ℔s. to the square inch), observing closely the part of mouth that is being operated upon; also watch the actions of the face and eyes. All these parts must be seen at the same time, a practice that cannot be successfully acquired without much experience.

Should the patient's throat begin to contract or his eyes close, or there occur any other sign that indicates contraction of the fauces, instantly cease throwing the spray, withdraw the instrument, and at the same time request him to take the tongue depressor out of the mouth. All this must be done, if possible, before he retches. At once request him to clear his throat. This act will rest him, as holding one's mouth open for two or three minutes is a little fatiguing. Then continue to make other application until all the remedy is thrown into the pharyngo-nasal cavity, keeping in mind these directions.

If the medicament has been made warm and thrown to the parts indicated, the patient will voluntarily say that he experiences a sensation of relief and smoothness in his throat; and if asked to locate the place where he feels the relief, he will at once place his fingers over and below the larynx; many of them, especially those who have a tickling cough in the larynx that has lasted for a few weeks, will place their hand on the upper portion of the chest.

Next make an application by means of the spray producer No. 5, applying with it the same remedies, and introducing the point of the instrument in the same way as with the No. 4, using the same precautions.

If the patient is not conscious that the stream from the instrument is going into his nostrils, it indicates that the mucous membrane is in quite an anæsthetic condition and that the inflammation is quite chronic, consequently it may be slow in yielding to the treatment.

Third, make an application to the fauces while the patient is slowly and deeply inhaling, using the spray producer No. 1. By this instrument the same vaseline compound is applied, with the addition of three to five drops of the following:

℞. Pinus Canadensis (Kennedy's)..............grs. iij.
    Glycerinæ (Price's)...............................ℨ ij.
    Acidi Carbolici.....................................ğr. ss.
    Ol. Galtheriæ.................................... ....gtts. v.
    Aquæ Ferv............ ....ℨ vj.
M.

This makes a pleasant mixture and has a very soothing effect on a throat and soft palate made sore by coughing.

I usually use five drops of the Pinus Cand. comp., during the first three treatments, then four drops for the next three. After this, using one and two at each succeeding treatment. If too much of this astringent is employed, the patient will complain of a slight soreness in the throat on swallowing.

Do not make an application to the nostrils in front, because the force of the air will be almost certain to excite sneezing, which is very undesirable. After some ten or twelve treatments, the excessive sensitiveness of the anterior nares may be reduced, then a very gentle spray of the vaseline and eucalyptus may be thrown into them.

The spray producer No. 5 is the most important instrument because it throws the remedy to the regions, the middle and superior turbinated processes, where the inflammation is most severe, this location being the site of origin of the primary disease, the chronic catarrhal inflammation. It is also the most difficult instrument to handle, because the stream of spray that issues from it is liable to strike the upper surface of the soft-palate and thus cause contraction of the faucal muscles.

If the pressure of the air that is used for making the spray is too great, the application may give rise to a paroxysm of sneezing, an effect that is quite undesirable; most patients can bear an air pressure of 7 ℔s to the square inch, but some can only endure 4 ℔s.

The effect of these applications, when judiciously made, will be very agreeable to the patient, *relieving him at once* of many of his most annoying symptoms.

Indeed, so marked has been the relief experienced by many of my patients, that some of them have made remarks as ungracious as the following : " I guess this attack will not be very bad anyhow."

ELECTRICITY.

This remedy is a valuable adjuvant, one that should be employed in every case, especially toward the latter part of the treatment. I have not been successful in the employment of the Faradic current, but with the galvanic current, the patient will at once state that he experiences beneficial effects.

It requires from 6 to 18 Leclanche cells to produce the desired effect. I apply the negative pole (cathode) to the lower end of the sternum, and the positive pole (anode) to the seventh cervical vertebra. * By this application all five of the special senses may be excited. The sense of taste is always excited if the application is properly made; it is known by a metallic taste being experienced in the throat and mouth, proving that these two organs are under the influence of this agent. An application that does not in this way excite the sense of taste is inefficient. The positive pole may sometimes be applied with marked benefit over the nose and cheeks, care being taken to employ just sufficient strength of current to be slightly felt.

---

*. This is called Central Galvanization by Beard and Rockwell.

The length of time that the electricity is applied should not exceed three minutes. One minute is the length of time that I usually employ it on new patients, watchfully lengthening the seance to the full time, three minutes. Instantly reversing the current for a few times, as well as interrupting it, is frequently productive of good results. I always make the applications after completing the local treatment with the spray producers.

CONSTITUTIONAL TREATMENT.

All of these patients are habitually constipated, and their renal secretions are usually scanty, besides this, they are in a condition of body that easily becomes exhausted, and many of them are in a state of continual weariness, therefore a laxative, a diuretic and a tonic are indicated. The bowels should be maintained quite open; at least two operations each day will be beneficial, for a week or two.

Ten grains of quinine should be taken each night on going to bed. This should be continued every night until the paroxysms are reduced to about 50 *per cent.* of their usual severity then seven to five grains may be a sufficient quantity. Many patients think that such doses would keep them awake, but in cases that were not severe, the contrary was the effect. I have sometimes added to the quinine, five grains of bromide of potassium with excellent results, obtaining refreshing sleep.

Special Hygiene of Pruritic Catarrh (*hay-fever, etc.*) *

It is as prepostrous to expect to even alleviate a patient afflicted with pruritic catarrh without strictly following the rules of hygiene, as it would be to maintain a ship dry with a leakage in its hull or a man sober while continually imbibing large quantities of alcoholic drinks.

PROTECTING THE HEAD. THE HAIR.

If a patient who has suffered from annual attacks of this complaint for about five years, and whose head perspires freely, should make the mistake of having his hair cut so short that it cannot be parted, he will soon learn, to his sorrow, that but little can be done to lessen the severity of his paroxysms, until his hair again grows. A cap may afford him some protection but because of its too frequent removal, it will not take the place of the lost hair. A properly constructed wig will come nearest in doing this.

WIGS, HEALTHFUL TO THE BALD-HEADED.

A large proportion of persons who are afflicted with pruritic catarrh are bald-headed, and the scalp of very many of them perspire profusely on the slightest exertion. With such, a very slight draught of air is sufficient to bring on a paroxysm of sneezing. An acquaintance, who had the misfortune to be quite

---

* Read before the St. Louis Medical Society, May 17th, 1884.

bald, informed me in 1872 that he cured himself of his "hay-fever" by wearing a wig. He had suffered from this complaint for a few years, and observed that he was most liable to sneeze when his head was bathed with perspiration. If at such times he wiped his head with a handkerchief that had been wet, it produced a cold, chilly sensation to his head, and always caused sneezing; if he used a warm handkerchief he did not sneeze. He had a relative who was a wig maker, and who advised him to wear a wig to prevent him from wiping his head so often. It took him some weeks torture by the disease before his pride—AN EXCEEDING FOOLISH ONE—gave way. He felt an improvement on the first day of wearing the wig and did not have an attack after that season. Of course he continues to wear the wig. Besides relieving him of his annual attacks of pruritic catarrh, he was relieved of headache also, a complaint that he had been subject to for years before his attack of "hay-fever."

I strongly urge all my bald-headed patients, whether afflicted with pruritic catarrh or with common chronic nasal catarrh, to wear a wig. The hair should be let grow until it is long enough to nearly touch the coat or dress collar; it should not at any time be much shorter or longer on *any person*, male or female.

The beard should be allowed to grow until it forms a good protection to the throat and neck. Shaving is a flagrant violation of one of the laws of health.

HATS AND CAPS.

The best hat for male patients is the soft hat.

A light skull cap should be worn day and night when the patient is in the house. It is not necessary to have a different cap for night wear, unless a warmer one was required at night, for the protection of the head is equally essential during all hours of the day and night.

All of these patients, male and female, perspire very freely about the head, and while the scalp is thus covered with moisture, even a slight draught of air will, in a few minutes, reduce the temperature of the surface fully 20°F. which in all probability, will be sufficient to produce a paroxysm. The cap is intended to prevent this sudden lowering of the temperature, not for the purpose of keeping the head warm.

Female patients should wear a silk hood day and night, it need not be very heavily quilted.

Those patients who do not require the inunction of the whole body with vaseline, may require to have the face, neck, hands and feet anointed with vaseline, as they retire for the night, as described in the section relating to local treatment.

CLOTHING.

Patients of both sexes should wear thin stocking-knit, cotton and wool mixed, vest and drawers, and a heavy suit of pure flannel over them. The advantage of wearing cotton next to the body, is that it absorbs the perspiration, thus preventing a cold, chilly sensation, should the body be exposed to a draught of air. Some of my patients have felt the necessity of wearing a third suit consisting of heavy flannel even on hot days, and claimed that they did not suffer in the least from excess of heat. This class of patients and all whose nasal passages are affected with catarrhal inflammation require a large amount of clothing and they bear it with great comfort.

INUNCTION OF THE BODY.

This is very frequently productive of marked benefit. The room in which the inunction is to be made should be kept quite comfortable. Vaseline is the substance to be used. It should be rubbed on by means of a flannel cloth made hot over a lamp. The clothing should be removed to the waist, and the body well rubbed, occupying about fifteen minutes time, then the clothing should be replaced, and that of the lower portion of the body removed, after which this part also should be well anointed, occupying about the same length of time. Some patients are remarkably fond of this operation and spend an hour and even longer in its performance.

THE FEET.

Male patients should wear boots, females high shoes. Two pairs of stockings should be worn; the pair next the feet should be cotton and the other woolen.

If slippers are to be worn, a pair of heavy woolen stockings should be drawn on over the stockings already on the feet.

Females must not wear elastic garters. In order to maintain the hose in place, they should be pulled on over the thin underdrawers, and held by four elastic straps, each of which has brass loops on each end, so formed as to securely retain the hold on the drawers and the top of the hose. In this way the circulation of the blood in the limbs is not impeded.

THE SLEEPING ROOM.

The sleeping room should be large, and well swept and dusted every day, it should face the South and East if possible. From morning until 1 P. M,, all the windows and doors should be left wide open; after that time they should be all closed up tight and the sunlight be excluded to almost total darkness. A piece of ice, weighing about 10 lbs., hung up about 6 feet high, in the middle of the room, will lower the temperature of the air to a pleasant coolness, and it will continue so during the whole night. Some might think that this would make the air of room too damp, but such is not the case.

If the ice melts too rapidly so that the air is made too cold, the ice may be covered with a piece of cotton or woolen cloth, with the woolen, the ice will melt more slowly than with the cotton covering. A swing to hold the ice may be made of a common towel, stretched and held by the four corners. This leaves the ice exposed to the downward current of warm air, which, as soon as it strikes the ice is lowered in temperature, continues in its course to the floor, forming the lower stratum of air in the room.

The water from the ice may be caught in a bucket
or other receptacle as it drops from the towel.

SLEEP.

The patient should sleep between blankets, but not
on feathers or old moss or old hair, a cotton mattress
is the best.   If a cotton mattress is not used then a
heavy cotton quilt should cover the bed mattress.   It
will be well to have the pillows made of cotton.

Annointing the face, neck, hands and feet with vaseline, just before retiring is quite refreshing, because it
is cooling.

The " catarrhal season " should be slept away if possible, but it is not best to sleep so much during the
day that the night will be passed in wakefulness.   If
the patient cannot sleep sufficiently long at night, an
anodyne should be given, but as a usual thing the ice
and quinine produces refreshing repose.

THE DIET.

A good, nourishing diet is advisable.   Everything
that the patient thinks that may disagree with him,
and all those articles knows to disagree with him,
should be avoided.   Going to bed very hungry may
prevent a good night's sleep.   Drinking water is
always healthful. One to two teacupfulls of hot water
as soon as the patient rises from bed in the morning,
or if convenient, before rising, is frequently conducive
to good digestion.   Milk, if taken after dinner, is liable to induce a cough by its causing the mucus in the
throat to become quite thick and adherent.

EXERCISE.

Many of these patients suffer from palpitation of the heart when they take exercise, but some gentle exertion, even to the extent of inducing a slight perspiration, is quite beneficial. As a general thing, the avoidance of sunlight, dust, smoke and other irritating agents that float in the air is the most conducive to comfort. Walking in a close, darkened room, in which a piece of ice is hung, to keep the temperature fully 10° F. to 20° F. below the outside temperature, is usually quite refreshing.

TO BE AVOIDED.

Sufferers from this complaint should not bathe; should not smoke, chew or snuff tobacco; should not drink beer, wine, whisky, brandy, gin or any beverage that contains alcohol; should not be out in the night air, should not allow themselves, under any circumstances, to become angry. The disease has a tendency to make one irritable, but this condition of mind must be controlled. A fit of anger will be almost certain to induce a fit of sneezing. Every victim of this complaint can, if he chooses, cultivate a habit of becoming angry, to his own discomfiture, or of exhibiting a disposition of patience. Coughing and sneezing must be avoided if possible. The former may many times be controlled to almost complete suppression. Handkerchiefs that have become wet from nasal secretions and tears, should be put out of the room. If the expectorations are very profuse, a spittoon filled with dry earth should be kept in the room and new earth put in it every morning.

SURGICAL TREATMENT.

Local medication and hygienic measure being completed, attention will now be given to harsher measures, which are to be employed in case the milder course has failed to give the desired relief. Before . giving the description of the surgical operations, I feel it my duty to make a few

REMARKS IN FAVOR OF CONSERVATIVE SURGERY.

Immediately successful results are always desirable in the healing art. Nothing captivates the physician and patient like an operation that promises immediate relief, by it days of uncertainty and suspense are to be banished and in their place, the victim is to quickly enjoy life free from torment. Especially is this result satisfactory in the highest degree, when the relief relates to a complaint that all our old and respected authorities have acknowledged was incurable. But with this grand advance in Rhinology, we must not be unmindful, that while the removal of diseased tissue is followed by marked relief, yet there is danger that this benificent operation may be employed too frequently, and in cases where it will be certain to do lasting harm.

Suppose a man, suffering from a violent toothache should visit one of our cheap " tooth-pulling " shops, and have it extracted, without first informing himself as to the possibility of its being saved for future valuable service; every one would at once pronounce him exceedingly foolish for thus impatiently sacrificing so important an organ. No doubt he would, in this way, be relieved of the pain in his tooth in a very much shorter time, than even an expeirenced dentist could have relieved him while saving his tooth, but what is to be said of the loss he has sustained? Is the loss of a tooth of so small a moment that it is not to be taken into consideration? In other words: is relief of the annoyance of an aching tooth in the quickest possible way the best under all circumstances? " Tooth-pullers" might say yes, but no dentist would.

In this dental case, we have almost a complete paralell to the operative method for the relief of pruritic catarrh. As there are numerous cases in which it is far better to extract an aching tooth, than try to save it, so are there numerous cases of pruritic catarrh in which it is far better to remove the offending membrane, than to endeavor to relieve the complaint by treating the originating inflammation; at the same time there are many aching teeth that should NOT be extracted, so are there many cases of pruritic catarrh that can be cured completly without any other treatment than the MILDEST measures for removing the chronic catarrhal inflammation of the nasal cavities. The mucous membrane of the nasal cavities, like the teeth, is a very valuable organ, and is essential to the welfare of the patient.

Let us retain all the benefits to be derived from surgical operations that will afford immediate relief of this complaint, but, at the same time, let us be careful not to entail a lasting injury by a rapid method,

when the same object can be attained by a slower
method that would preserve the mucous membrane
for future use.

I believe that every case of pruritic catarrh can be
relieved of the pruritic symptoms at once by the
"destructive process", but I do KNOW from experience,
that many cases can be cured without desroying the
mucous membrane—without committing irreparable
damage to this important organ, the loss of which
may cause the patient to lead a life of unhappiness, if
not of torment from an ozenic condition of the nasal
cavities, that sometimes follows such operations.

I do not wish to act the part of an alarmist, but desire to give some wholsome warning, that I am certain will be beneficial to the profession and their patients.

IMPORTANT FACTS.

There are several very important facts that should
engage the careful considertion of every physician
who contemplates performing an operation on the
nasal cavities, namely: 1st., a cicatrix follows every
application of the galvano-cautery, nitric acid, chromic acid, etc; 2nd., this scar membrane is not mucous
membrane; 3d., this scar surface is always dry; 4th.
the nasal secretion flowing from mucous surfaces superior to it is certain to lodge on this dry spot, and
there become inspissated and be the occasion of several very annoying symptoms.

Before inspissation takes place, the mass becomes
decomposed, and if the muco-purulent secretion is

profuse, it will affect the breath unpleasantly, a very serious matter if the patient is a young woman. With the decomposition, it acquires an acid property, which will cause so much irritation at the place of lodgement, that the patient will not desist picking and blowing the nose to free himself of it. The blowing not unfrequently resulting in ear complications, the secretion in the pharyngo-nasal cavity being forced up the Eustachian tube and into the tympanum, where it may originate tubal and middle ear catarrh.

I have seen, during the spring of this year (1885) five persons who had the formation of crusts in the nostrils none of whom had such formations previous to their being operated upon by the galvano-cautery. Such cases are very much more difficult to cure than the so-called atrophic catarrh. Indeed three of these cases were pronounced atrophic catarrh by a physician previous to their calling on me.

I account for the formation of this inspissated collection in this way: The scar tissue, that follows all destructive applications, cannot perform the functions of mucous membrane, as it has not a single mucous gland in it, consequently IT MUST ALWAYS REMAIN DRY, unless moistened by contiguous mucous membrane. Being always dry, secretion that lodges there, remains there, the heat of the cavity evaporating its watery portion, causing inspissation.

The point I wish to make is this: THE LESS SCAR TISSUE THERE IS FORMED IN THE NASAL CAVITIES, THE BETTER FOR THE FUTURE WELFARE OF THE PATIENT.

If it is found upon trial that the treatment for chronic catarrhal inflammation will not relieve the patient of the pruritic nasal symptoms, THEN OPERATE, BUT NOT BEFORE; no physician is justified in resorting to severe treatment before he has proved that a milder course has been ineffectual.

THE KIND OF CASES TO BE OPERATED UPON.

As indicated in the closing paragraphs of the last chapter, page 106, only the most chronic cases, principally those affected with the middle and late forms, will require surgical treatment. I have not yet operated on a case of the May and June form, and quite a number of the July and Autumnal forms have recovwithout surgical interference.

My method of ascertaining who will require operative procedure, is to treat by the spray producers EVERY CASE for a few days, giving from five to fifteen treatment. From the effect of these applications, I judge whether or not the case will require severer measures. It is seen that I operate on a few patients as possible because I fear the effects that will follow the formation of scar tissue in the nasal cavities. I do not wait until the pruritic season as passed away, but operate as soon as I find that the spray producers will not cure the complaint.

RELIEF BY SURGICAL MEASURES.

This consists in the removal of the diseased, hyperæstheic mucous membrane that covers the turbinated processes and portions of the septum nasi. This is done by means of Jarvis's wire snare, or by caustics, such as chromic, acetic or nitric acid, or by the galvano-cautery. The galvano-cautery and Jarvis's snare are considered the most reliable. Dr. J. A. Stucky of Lexington, Ky. uses, with excellent results, chromic acid. I have grasped the sensitive portions of the membrane with a slender, but strong pair of forceps, maintaining the hold on the membrane for two or three minutes, first spraying the parts with a 2 *per cent.* solution of cocaine. The result was quite satisfactory.

LOCATING THE DISEASED MEMBRANE.

In locating the hyperæsthetic spot or spots, I employ, if possible, a small reflector (Fig 4.) if it can

Fig. 4.

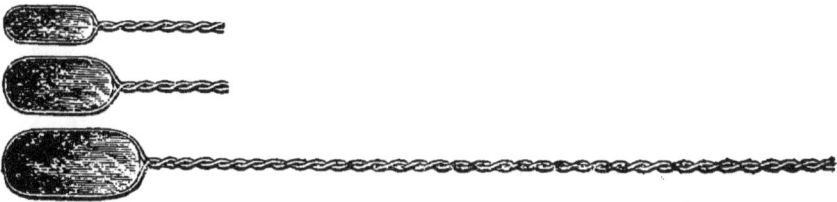

ANTERIOR NASAL MIRRORS.—The mirrors are represented full size The handles are five inches long. The desired angle may be given to each mirror by bending the wire handle near the glass.

be passed into the anterior nares without producing much if any irritation, using, at the same, a nasal speculum. I then insert a probe, bent slightly at the point, and ascertain according to the method employed by Dr. Roe, of Rochester N. Y. (see appendix) the location of the most sensitive spot known by the patient experiencing a BURNING SENSATION. This spot is then touched, after it has been anæsthetized by cocaine. This valuable anæsthetic renders these surgical procedures comparatively painless during the time of the operation, but does not prevent considerable pain for several hours afterward.

APPLICATION OF THE GALVANO-CAUTERY.

Every one has a favorite manner of applying the galvano-cautery. Some allow the platinum to become almost white hot before passing it into the nasal cavity. I did this on several occassions to my patients great detriment, the radiating heat being so great as to cause acute inflammation of the whole cavity and great swelling of the face. The electrical energy should be sufficient to make the platinum white-hot IN ONE SECOND of time. Of course if the current was allowed to continue, the wire would be burnt in about three or four seconds, but when the electrode is laid on the tissue, this keeps the wire from becoming sufficiently hot to be destroyed.

I prefer to place the electrode ON THE SPOT to be cauterized, and then make the connection with my foot, never using my finger or thumb for making connection, as this would necessitate holding the instrument so firmly in my hand that I could not be certain of the degree of pressure I was making on the part being burned.

Immediately on the withdrawal of the electrode, I spray the cavity with spray producers Nos. 2 and 5 (see page 111.) employing the vaseline comp. given on page 111. This will have a soothing effect, but if the patient still complains of the distress from the burning, apply the oleate of cocaine. This is an ex-

-cellent preparation, and produces a much more lasting effect than the solution. The strength I now employ is 5 *per cent.*

The next day, the patient should receive the regular treatment with the spray producers. Usually the first application of the electricity has a marked effect, reducing the pruritic symptoms.

As soon as the patient can endure a second application of the cautery, it should be applied. Generally one and two applications a week can be borne without great discomfort.

Constitutional treatment should be given while the local measures are being employed.

JARVIS'S SNARE.

I prefer this instrument to the galvano-cautery, because it can be employed to remove even an extensive hypertrophy of the turbinated processes without leaving a large cicatrix, certainly not the one-tenth of the size of the portion removed.

After engaging the loop of the snare around the hypertrophy, I tighten the nut sufficiently to prove that the snare has a hold. I then direct the patient to take hold of the instrument with his left hand and turn the nut with his right hand. This he should do so as to cause but little pain. It generally takes about half to three-quarters of an hour to cut off a large hypertrophy. When the instrument has cut itself out, if the patient does not blow his nose—which he should not do—the removal will be made without the loss of blood, or at least with very little; but best of all, the size of the scar left is about the size of a large pin's head. Of course, cocaine should be applied before the snare is adjusted.

This method of removing diseased tissue pleases me very much. I feel like thanking Dr. Jarvis every time I use his instrument, as with it I preserve my patient from the evil effects of a large cicatrix.

CHROMIC ACID.

I have used this acid a few times, but probably did not apply it just in the right manner. The reader will be pleased with the success that Dr. J. A. Stucky, of Lexington, Ky., has had with this powerful agent. His views and experience may be seen in full in the appendix.

One of my patients, a delicate lady, was so severely affected by pain in her left ear, occasioned by the effect of chromic acid applied in the left nasal cavity, that I was obliged to perforate the membrana tympani to give her relief.

I applied the chromic acid by means of a probe. I heated the point of the probe and touched it to a crystal of the acid, which instantly melted and coated the probe. I then touched the hypertrophied membrane with the probe, aided by the anterior nares reflector inserted into the cavity from the front. Two applications had been made, but these did not occasion very great disturbance. I may have held the acid on the parts too long, and so produced a more lasting impression than was required. A 2 *per cent.* solution of cocaine was applied two separate times before the acid was used.

NITRIC ACID.

I employed nitric acid but one time. The disturbance occasioned by its application was so great that I think that I will not use it again.

POSTERIOR NARES.

If the posterior portions of the turbinated processes or the septum nasi are to be inspected or operated upon, I hand the patient the tongue depressor (Fig. 1) and direct him to hold his tongue down with it, using his left hand. If the space between the posterior wall of the pharynx and the soft palate is sufficiently large, I place the pharyngeal reflector (Fig. 5.) back in the fauces to get a reflection of the

Fig. 5.

PHARYNGEAL MIRROR.--By pressure on the level on the handle the mirror may be made to take any desired angle, thus reflecting the posterior, superior and anterior surfaces of the pharyngo-nasal cavity, while rotation on its axis reflects the lateral surfaces.

posterior extremities of the inferior and middle turbinated processes and the septum nasi, using my left hand, leaving the right hand for the manipulation of the diagnostic probe, the electrode, the Jarvis snare, or the chromic acid probe.

If the velum hangs too close to the posterior wall of the pharynx, I hook the pendent portion with the spreading soft palate retractor (Fig. 6.). Before

Fig. 6.

SOFT PALATE RETRACTOR.—A, Lever to seperate the arms. B, The soft rubber band that closes the arms and holds the uvula out of the operator's way. C, The lever that raises the wedge. After the instrument is is introduced behind the velum and the arms spread by the lever A, then then the wedge retains them in position.

. drawing the palate forward, I spread the limbs of the instrument a little, and then draw it slightly outward. I then lift the right hand of the patient to the handle of the instrument and direct him to draw it as far forward as he can without causing unpleasant sensations. The patient can hold the instrument very much better than an assistant, as he knows how to control it so that it will not cause him to retch or occasion pain.

The reflection from the pharyngeal mirror will assist the operator in locating the sensative spots, and in adjusting the Jarvis snare.

In making all caustic applications to these parts I employ the same methods.

APPENDIX.

As the most of this little book was written one, two and three years ago, I could not embodie, without rewriting each chapter, the opinions of many of the late writers on this subject; but as the views of many of them are important and the information very valuable, I have concluded to place them all together, in the form of an appendix, giving those first that were published first.

EXTRACTS FROM DR. WM. H. DALY'S ARTICLE.

Dr. Wm. H. Daly of Pittsburg, Pa. read a paper before the American Laryngological Association in 1881, and published in the Archives of Laryngology, April, 1882, " On the Relation of Hay Asthma and Chronic Naso-pharyngeal Catarrh."
He says:—

" Let me ask that patients suffering from hay asthma be thoroughly inspected, and if evidence of chronic inflammatory naso-pharyngeal disease and its resultant derangement of sensibility and secretion exists, treat it, losing sight, as far as possible, of the exciting or atmospheric cause, that can in no wise be removed without removing the patient.

"If hypertrophic enlargments of the mucous membrane in any part of its distribution are found, reduce them or remove them entirely.

"If polypi or polypoid growths are found, remove them, cauterizing thoroughly the bases of the growths.

"If there is chronic disease of any kind whatsoever, put the parts in order, and thereby enable them to withstand the exciting influence of the next recurring crop of bacteria. A preternaturally irritable condition of the parts, from any disease whatsoever, will render them liable to respond to the effect of influences that would be entirely innocuous if applied to healthful tissues. If we resort the diseased tissues to a healthful condition, and hay asthma recurs, then we are warranted in considering the individual case a neurosis, and not otherwise.

"The histories of the following cases show beyond doubt that the annually recurring attacks of hay asthma were not so much the result of the mere presence of bacteria in the atmosphere as the fact that these bacteria had their peculiar effects upon parts rendered susceptible to their irritating influences by chronic local disease of the naso-pharynx. "

He then gives the history of several very interesting cases, the result of whose treatment fully substantiates all that has asserted in the above five paragraphs

His first case consulted him in December 1878, and found that he had hypertrophic growths.

"* * Destruction of the growths was accomplished with the galvano-cautery, and treatment during the next three months resulted in a great diminution of the hypersensibility of the nasal mucous membrane. The question was later asked, as to what would be the prospect of immunity from the expected attack of hay asthma. While giving him no encouragement, I advised him to wait and see. The dreaded 15th of June, 1879, was passed at home. The summer and autumn passed also, and the spring and summer of 1880—81 were all passed at home, without any recurrence of hay asthma."

He closes his interesting and valuable paper with three following the conclusions:

"1st. That in a proportion of cases there is an intrinsic condition of local chronic disease upon which the exciting cause acts with effect.

" 2d. Without this intrinsic local disease the exciting cause is innocuous.

" 3d. The patients believed they were only slightly if at all, affected with naso-pharyngeal disease of a chronic character."

Extracts from Dr. Roe's Paper.

Dr. J. O. Roe of Rochester, N. Y. has been successful in treating this ailment by surgical means. In
the fall of 1879 he operated on a patient of his, who
"had been a sufferer from hay-fever at least twelve
years." Dr. Roe removed the turbinated hypertrophy with a galvano-cautery.

In speaking of the result of this operation he says :

"Greatly to his, and also to my surprise, he was not
attacked by hay-fever during the following summer;
and, as he informed me a short time ago, he has been
entirely exempt from it during each summer since."

From the following quotations, taken from Dr.
Roe's second article, read before the Medical Society
of the State of New York, Feb. 1884, his method of
procedure is seen :

"In the treatment of hay-fever we should first determine, by a careful exploration of the nasal chambers,
the exact nature of the conditions which have been
the exciting cause of the hyperæsthesia. Each parti-
cular spot which is especially sensitive should be located, and receive thorough and careful treatment until
this sensitiveness is removed and no sensation of hay-
fever is experienced by the patient when these regions
are touched. This hay-fever sensation is unmistakable
by the patient, for on touching these regions, however
lightly, a burning sensation is felt in the nostril, as if
the probe were heated, and is attended by the usual
reflex phenomena.

" When hypertrophied turbinated corpora cavernosa are the seat of the sensitive region, they should

be throughly removed. When this region is the seat
of the sensitiveness, though there is no well-marked
hypertrophy of the turbinated bodies, sufficient tissue
should be removed to distroy the diseased and sensitive terminal nerve filaments and to obliterate the enlarged blood-vessels. Redundant and hypertrophied
tissue is best removed with Jarvis's snare, although
caustics, such as acetic, chromic, or nitric acid, may be
employed. For the distruction of the deeper plexuses
of vessels, the galvanic cautery is by far the most efficient. It is also the most efficient means of removing
the sensitive regions on the septum and other portions
of the nasal chambers. For the latter purpose, a very
small point should be used, so as to enable the operator
to limit the cauterization entirely to the diseased tissue,
and, by using a very small point, but little pain is occasioned.

"All obstructions to the nostrils other than hypertrophic tissue should be removed, and also all abnormal conditions of the passages, whether they be sufficient to cause obstruction to the chambers or not,
should be corrected.

"In all these cases it is of special importance that
there should be no points of contact between the turbinated bones themselves or the turbinated bones and
the septum, even though there be no obstruction
whatever to respiration. Spiculæ of bone are often
found projecting across like a spur and exciting irritation and producing thickening of the opposite surface. This condition is more often found between the
middle and superior than the inferior turbinated bone
and the septum.

"Afterward, when all offending tissue has been
removed, local medication should be made to the
nasal passages until the parts are healed and the

chronic rhinitis cured, and the special irritability and hyperæsthesia has disappeared from every portion that is shown by the exploration with a probe to be abnormally sensative.

"The time when these radically curative measures should be instituted is, my observations lead me to believe, when the patient is free from the affection, and in time to allow thorough healing of the parts before the time of the expected attack, although, if necessary, it may be begun during the attack.

"It is also advisable and even necessary (where there is a doubt as to the sufficiency of the treatment) to examine the patient from time to time during the hay-fever season to observe if any portion of the nasal mucous membrane becomes irritated that has before been overlooked. If so, it should then receive prompt attention, and the diseased portion be thoroughly removed.

"The practical outcome or result of this method of dealing with hay-fever is, after all, the most interesting evidence as to its value. Of the five cases which I reported to the society last year, four have been accessible, so that I have been able to determine the result in these."

The following case will be read with interest and instruction:

"Case VI.—J. R., aged twenty-seven, a stout, well-developed man, was referred to me March 9, 1883. He had had hay-fever for eight years very severely, his attacks coming on about August 10th and continuing until frost came, and being attended with more or less asthma. During the remainder of the year he had more or less catarrh and frequent colds in the head, with marked stoppage of the nostrils.

When free from colds, his nostrils were clear and un-
obstructed. The inhalation of dust or any marked
irritant at any time would cause sneezing and temporary stoppage of them.

"Examination showed moderate general thickening of the mucous membrane of both nasal passages,
but the turbinated bodies were not noticeably hypertrophied, nor were there any bony obstructions. On
exploration marked sensitiveness was found all along
the inferior and middle turbinated bones, especially
at the posterior end, giving rise to the' characteristic
sensations of hay-fever. A similar sensitive region
was found along the lower portion of the septum on
both sides, and on the left side it was also very sensitive along its upper portion.

" *Treatment.*—The sensitive turbinated tissue was
cauterized sufficiently to destroy the hyperæsthesia
and to obliterate the enlarged vessels which the frequent, sudden and great swelling of this tissue indicated to be the case. A very small point was used,
so as to give the least amount of pain. The sensitive
organs of the septum were also touched with the cau-
tery-point. Afterward local treatment was continued
to the passages for three or four weeks, until the
parts had healed and no symptoms of hay-fever could
be excited in any portion of the nasal cavity.

"He was travelling most of the time, and during the
latter part of August went West. In November I
heard from him that he had escaped entirely his
annual attack, although he was in the region where
others were having it and where he had had it
before."

PRURITIC RHINITIS BY P. W. LOGAN, M. D., PRESIDENT
OF THE AMERICAN RHINOLOGICAL ASSOCIA-
TION, ETC.

Knowing that Dr. P. W. Logan of Knoxville,
Tenn. has had quite a number of patients (I call fifteen or twenty quite a number) affected with pruritic rhinitis under his care during the last three years.
I wrote to him asking his views, experience, etc.
It is seen that he corroborates what I have given in
this work. The following is his reply received May,
22, 1885:

DR. RUMBOLD.—In response to your letter asking
for a report of my hay-fever cases, I must say that
my experience in the treatment of this complaint has
not been extensive. So far as I am able to judge however, I believe that hay-fever is mainly due to a peculiar excitement of the nerves of the parts which is
liable to manifest itself by the symptoms known as
hay-fever, June catarrh, rose catarrh, peach cold, summer catarrh, autumnal catarrh, pruritic rhinitis, pruritic catarrh, etc.

It is singular that this affection should manifest itself
in some instances with a periodicity so marked and
precise, as to appear on the same day of a certain
month each year; especially when it is known that
pollen may not appear exactly at the same time every
year. Our seasons being sometimes early and at
other times late.

As pertaines to the various and numerous exciting
causes, dwelt upon by several authors, I wish to speak
more especially of pre-existing nasal disease, as an
important factor in the development of this com-

plaint. I refer to chronic inflammation of the nasal mucous membrane and some of its sequences such as thickened turbinated processes, growths· etc. In all of the cases of pruritic catarrh examined by me, I have found pre-existing nasal inflammation, or positive evidence of its existence before the development of the complaint. Nasal inflammation however, would not in my judgement give rise to this complaint except in a patient possessed of this peculiar idiosyncrasy to which the disease is remotely due.

In some cases occurring during the winter months or continuing into winter, I have been unable to ascertain any other exciting cause than the presence of a catarrhal inflammation. As positive evidence of the fact that hay-fever may owe its existence to the presence of catarrhal inflammation or growths, relief from this inflammation, etc., will in some instances, I am sure, bring permanent relief from the usual attacks. This seems to me positive evidence of the fact that nasal trouble occurring in a subject possessed of a predisposition to hay-fever is an important exciting cause which should not be overlooked in the etiology of this disease.

Like yourself, Daly, Roe and Hack, I am inclined to the opinion that catarrhal inflammation of the nasal mucous membrane exists in many if not in all hay-fever victims, and in order to cure the ailment, we must relieve the nasal trouble by a mild, soothing and unirritating application of vaseline and oil of eucalyptus or oil of winter-green $\mathfrak{z}j$ of former to gtt, v of latter, very gently sprayed behind the soft palate, into the vault of the pharynx and posterior nares, and over the lower pharynx. At the same time removal of growth in the nasal passages or the thickened and sensitive mucous membrane covering the turbinated processes is necessary.

Where it is possible, I have found it best to treat the pre-existing nasal disease anterior to the time at which hay-fever usually appears. I can recall several cases whom I treated for catarrhal trouble during the spring months preceeding the time for the usual attack, who after taking treatment did not have a return.

The internal administration of quinine and other tonics, diuretics and laxative, I have found very beneficial in connection with local treatment, but have failed to get any perceptible beneficial results from valerianate of zinc and assafetida so highly spoken of by Dr. Mackenzie of London.

Hay-fever patients certainly need a tonic and sustaining course of treatment, hence the good effects of mountain and sea air. I believe that attitude and sea air counteract the operation of pollen, thereby relieving many cases of this disease, yet there are exceptions to this general rule. While pollen may be, and no doubt in prolific in exciting an attack of this malady, I do not think it is always due to this cause. I have had a few cases who were very easily effected by dust of any kind and by light. I treated them locally with the following: Vaseline ʒj, oil of wintergreen gtt v. This mixture was applied with the spray producers Nos. 4, 5 and 1, to the vault of pharynx, posterior nares and lower pharynx. The use of astringents in these cases I have found injurious therefore discard them.

Why the predisposing and exciting causes of this disease should exist in a greater degree in England and America than in any other countries, I am at a loss to understand, as we take cold everywhere and

are exposed to pollen, dust, etc, in France, Germany
and other countries, where hay-fever is said to exist
to a very limited extent if at all. While it is said to
occur less frequently in the Southern than in the
Northern or Western portion of the United States,
catarrhal troubles are likewise not so common, nor so
inveterate, so far as their succesful treatment is concerned, in the South as in the North. Climate alone
however is inadequate to cure catarrhal inflammation
of the upper respiratory tract. As constitutional predisposition or idiosyncrasy seems necessary to the es-
tablishment of hay-fever, we can only counteract development of the same by subduing rhinal inflammation
and removing other conditions existing in the nasal
passages, which might give rise to development of
this disease. Hay-fever patients unless treated, grow
worse from year to year. The same is true of catarrhal patients generally.

In making local applications to my hay-fever patients I soon learned that they should be treated very
gently. The spray should be used with as little air as
possible to produce a spray sufficient to gently cover
the affected surfaces, and the patient must not be treated too often. As the mucous membrane improves,
applications should be repeated less frequently.

I have experienced better results, so far as local
applications are concerned, from vaseline and oil of
eucalyptus, applied in the form of spray—than from
any other remedy or combination of remedies. Of
course I directed my patients to guard against the operation of everything which might excite or aggravate
an attack of their trouble.

                    P. W.Logan, M. D., Knoxville, Tenn.

PRURITIC CATARRH OR HAY-FEVER.—ITS TREATMENT.*
By J. A. STUCKY, M. D., of Lexington, Ky ; Sur-
geon to St. Joseph's Hospital; Member of the
Kentucky State Medical Society ; Vice Presi-
dent AMERICAN RHINOLOGICAL ASSOCIA-
TION, etc.

As Dr. J. A. Stucky read a paper on this subject
before the American Rhinological Association in October, 1884; I wrote to him asking a copy for this
work which he has kindly sent to me.

I am sure that every physician who will read it
through, will be struck with the thoroughness in
which he handles the subject, and every interested
reader will thank him for the very valuable information he has given.

The following is his contribution to the Association :

In calling your attention to the subject of Pruritic Catarrh or so-called Hay-Fever, I am not unmindful of the fact that I am but adding to an already
over filled list of contributors to the literature of this
subject.

The many conflicting theories as to the etiology
and pathology of this disease, stimulates me to contribute my mite. Probably no disease of the supe-

* Read before the American Rhinological Association,
Oct., 1884.

rior respiratory tract causes more suffering than so-
called Hay-Fever. I shall not enter into a detailed
description of the symptoms, etiology or pathology
of the disease but am led to offer a few suggestions
as to the treatment, because of results obtained. In
the beginning, I desire to enter a protest against the
term "hay-fever" and all other names of a similar
nature, giving as a reason, that the terms used to designate the disease in question are meaningless, hence
not entitled to the respect they now have. You are
doubtless aware of the fact that I am not alone in
taking this view, but am simply following in the wake
of Dr. Thos. F. Rumbold and joining with him in the
plea for a name more expressive, appropriate and
descriptive of the disease.

If the odor of hay, roses, etc., were the only
known causes of this disease, we would be satisfied
with the time honored names *hay-fever, rose-cold* etc.,
because they would indicate its nature.

In a recent article on this subject, Dr. Rumbold
says: "If the name indicate that a certain prominent
fact or feature of a disease is constantly present so
as to distinguish it from other diseases, when such *is
not* the case, then most certainly the misguiding name
should be discarded; as its retention will be very
liable to lead to an erroneous diagnosis, and thus a
case might be excluded from its proper class, and, as
a consequence, be improperly treated". Dr. G. M.
Beard, a leading authority, Admits the substantial
identity of Autumnal Catarrh and June Cold, etc.

"The inappropriateness or rather insufficiency of
the term hay-fever is now quite generally admitted;
for even where the predisposition exists, hay of any
kind, fresh or dried, acts as an exciting cause in but
a minority of cases, and rarely, if ever, is it the only

irritant that gives rise to the paroxysms." * "Yet the phenomena of the disease are alike in all cases whether they occur in the Spring, Summer, Fall or Winter."†

Without discussing this part of the subject at any greater length, I accept the name, suggested by Dr. Rumbold, as being the *most* appropriate because most descriptive of the disease.

I quote at length from a recent article by him:

"Pruritic Rhinitis or Itching Nasal Catarrh, is the name selected for this, as yet unexplained, phenomenon. This name is descriptive of its most prominent characteristics namely: *itching, inflammation* and *flow of mucus.* The attack is ushered in by an itching of the nose and face; this soon affects the eyes, causing intense suffering. The itching sensation in the nostrils gives rise to prolonged sneezing, this in turn makes the eyes worse; the itching soon reaches the soft palate and fauces, and to relieve these parts of the same sensation, the tongue is used to rub them. As the tickling is not relieved, a rasping cough is tried which is so persistently continued that the throat soon becomes sore, and, in older sufferers, shortness of breathing ensues, and symptoms of asthma are developed.   *   *   *

"Because of the uniformity of this symptom—*itching*—and the fact that it is always accompanied by inflammation, the name suggested indicates the *first*, the *principle* and most *prominent* symptom, which is characteristic of the malady at whatever season of the year the victim is attacked, and it is by no means misleading".

---

* Beard, on Hay-Fever, 1876.
† Rumbold. St. Louis Medical and Surgical Journal, June, 1884.

I have already referred to the numerous causes (?) of this disease, which I shall hereafter designate as *Pruritic Catarrh.*

It matters not what the exciting cause may be, whether pollen of hay, parasites—animal or vegetable—bacilli, odor of rose, inhalation of dust, or what not, the fact I have been able to demonstrate in every case I have seen, is this, that the disease is preceded by a nasal catarrh, and relief of the catarrh was relief of the "hay-fever." In every case examined by me, I have found the middle and inferior turbinated bone covered with hypertrophied mucous membrane; and during a paroxysm, the nasal cavities were completely closed by the swollen membrane, giving rise to the uncomfortable feeling and frequently excruciating pain in the eyes, cheek, frontal region, and in a few cases intense pain in the back of the head. A brief glance at the anatomy of the parts will easily convince us how that by completely closing the openings that lead from the antrum of Highmore, the frontal, sphenoidal and ethmoidal sinuses, as well as the lachrymal canals, will give rise to all of the symptoms referred to. In addition to closing of these cavities, the inflammation extends into them in some cases, giving rise to alarming symptoms.

The membranous lining of these cavities is similar to that of the nasal fossæ. Whenever the nasal mucous membrane is in an active state of inflammation, which is attended with a great degree of swelling, the communication with the cavities mentioned must be shut off, and the accumulating fluids press against their boundaries, and as the pressure increases the pain becomes more intense. This pain ceases as soon as the imprisoned fluid finds an exit through its natural passage.

In the *American Journal of the Medical Sciences,*

July, 1883, Dr. Mackenzie, of Baltimore, Md., says:

"(1.) That in the nose there exists a well defined, sensitive area, whose stimulation, either through a local pathological process, or through the action of an irritant introduced from without, is capable of producing an excitation which finds its expression in a reflex act, or in a series of reflected phenomena.

"(2.) That this sensitive area corresponds, in all probability, with that portion of nasal mucous membrane which covers the turbinated corpora cavernosa.

"(3.) That reflex cough (or asthma) is produced only by stimulation of this area, and is only exceptionally evoked when the irritant is applied to other portions of the nasal chamber.

"(4.) That all parts of this area are not equally capable of generating the reflex act, the most sensitive spot being probably represented by that portion of the *membrane* which clothes the posterior extremity of the inferior turbinated body and that of the septum immediately opposite.

"(5.) That the tendency to *reflex action varies in different individuals*, and is dependent upon the varying degree of *excitability* of the erectile tissue. In some the slightest touch is sufficient to excite it, in others chronic hyperæmia or hypertrophy of the cavernous bodies seems to evoke it by constant irritation of the reflex centres, as occurs in similar conditions of other erectile organs, as, for example, the clitoris.

"(6.) That this exaggerated or disordered functional activity of the area may possibly throw some light on the physiological destiny of the erectile bodies. Among other properties which they possess, may they not act as sentinels to guard the lower air pas-

sages and pharynx against the entrance of foreign
bodies, noxious inhalations, and other injurious agents,
to which they might otherwise be exposed?"

The physiological phenomena referred to by the
author, are "to be found in the doctrine of correlated
areas, the reflex taking place through the vaso-dilator
nerves from the superior cervical ganglion of the
sympathetic."

In the beginning of an attack of Pruritic Catarrh
the first symptom is intense itching of the nose accompanied by sneezing. Gradually the nasal fossæ
fill up, until nasal respiration is entirely prevented.
When the disease has reached the point of occluding
the nares, by infiltration and inflammation of the
Schneiderian membrane, then the frontal and occipital
ache, with pain in the cheeks, accompanied by alternate chilliness and heat and a feeling of general discomfort with loss of appetite, pyrexia and general
malaise. Frequently do we meet with cases of inflammation of the middle ear that has extended
through the Eustachian tube, following attacks of
Pruritic Catarrh. I need not remind you that during
a paroxysm, the senses of taste, smell and hearing
are much impaired, and soon the sufferer becomes
prostrated. 1 shall not consume any of your valuable time by giving a description of the symptoms,
with the history of the cases I have treated. Will
only say, that of all the cases I have recorded every
one presented *all* or a *majority* of the prominent
symptoms characteristic of the disease. In conclud-
ing this part of my subject I agree with Dr. Harri-
son Allen that there is ' nothing peculiar to the disease, save its sharply defined periodicity, particularly
in that phase of it where the periods of recurrence
happen to coincide with the time of the fruitage of

certain plants, or the gathering of certain crops."

I now ask your attention to a consideration of the
method of treatment employed for this class of sufferers. If, as we have stated, the disease is due to a
nasal catarrh, and the paroxysms—asthmatic—are
brought on by an *irritant* of some kind, then the log-
ical inference is, get rid of the catarrh and you get
rid of the sequelæ, of which Pruritic Catarrh (hay-
fever) is the most annoying—for where the soil is in
a proper condition and seed sown, it sprouts sponta-
neously. " *Remove the cause, and cure the disease.*"

During a paroxysm our treatment is palliative only.
That we can in every case palliate and cut short the
aggravating symptoms I have no doubt. Experience
has demonstrated this. The treatment employed by
me is that used by Dr. Rumbold, with some modifi-
cations. If a patient presents himself during an attack I gently apply, by means of the spray, one-half
drachm of the following mixture.

℞   Acid Carbol...gr. j.
    Ol. Eucalyptol (Saunders).....................ℳ ij.
    Boric Acid...grs. x to xx.
    Glycerine..ʒj.
    Vaseline.......... ....ʒvij.
M.

One-half dram of this mixture is placed in the
bowl of the spray tube, heated and applied, by means
of compressed air, gently and thoroughly to the entire
pharyngo-nasal and post nasal cavity. The relief in
a majority of cases is almost immediate. The spray
tubes used in this treatment are those invented by
Dr. Rumbold, and in them we have a perfect instrument for thoroughly cleansing and medicating the
entire superior respiratory tract.

In a few cases it is necessary to omit the eucalyptol
on account of its stimulating qualities, when this is

necessary I use the same mixture with this exception. The properties of boracic acid need no eulogizing, they are well known, as antiseptic, slightly stimulating and soothing. Strong astringents or irritants of any kind not only do harm in this variety of nasal trouble *in the acute stage* but in all varieties. Our treatment should always be soothing, avoiding everything that irritates, this applies only to local medical treatment and not to the surgical.

I not only use boracic acid in this form of nasal trouble, but in others, principally the atrophic, in strength varying from 5 to 30 grs to the ounce, always using vaseline as the excipient and applying by means of spray. The application of this remedy in the manner suggested is made daily for a week, then twice a week until the disease is checked and symptoms disappear. Usually after the first or second treatment, patients express themselves as being relieved. This treatment suggested has given me excellent results in mild forms of this complaint in adults and still better results in treatment of children and persons under 25 years of age. The way in which the remedies relieve I suppose to be due to the fact that they remove, destroy or render inert the irritant that is imbedded in the mucous membrane, and by their soothing action reduce the inflammation and thus relieve the pent up secretions in the sinuses and cavities connected with the nasal organ.

I do not pretend to say what the exciting cause of this disease is, what causes it in one will not in another. The one fact I desire to emphasize is, the primary cause in every case is nasal catarrh. After relieving a case of the most prominent and distressing symptoms, and nasal respiration is thoroughly established, the inflammation and infiltration all gone, an examination with the rhinoscope (posteriorly) will reveal a mass—varying in size from a small buck-shot to the size of a hazel nut—of hypertrophied tissues over the posterior extremity of the inferior or middle turbinated bone, more frequently the former.

The treatmen already suggested, I stated, was palliative, and I desire now to speak of the radical treatment, which consists in the removal of the hypertrophied tissue. This is accomplished by means of the Jarvis Snare or, more frequently and preferably, by chromic acid. In the use of this agent my experience coincides with that of Dr. F. Donaldson, of Baltimore, Md., who says. "We have found chromic acid a powerful escharotic, not causing pain or hemorrhage and, when cautiously used, perfectly under control."

Its action is that of a prompt solvent of organic matter. It rapidly oxidizes and decomposes the tissues. It loses one half of its oxygen, and is itself converted into the inert sesquioxide. It is, at the same time, an antiseptic, and disinfectant. It appears, according to Wood and Bache, "to owe its antiseptic action to its power of coagulating albumen and all protean compounds, in which it has been found to exceed all the acids and metallic salts that have been tried, being ten times stronger than carbolic acid, fifteen times stronger than nitric acid, and twenty times stronger than bi-chloride of mercury." *It gives less pain than other caustics.*

It is one of the most powerful destructive agents to inferior organic life, greatly exceeding carbolic acid in this respect. The method by which we apply the chromic acid (*paste* made by adding just enough water to render it semi-solid) is to first dry the parts

with absorbent cotton wrapped around a nasal probe,
this should be done very gently so as to not excite
sneezing or cause pain? If the drying process causes
either it should be discontinued. If the application
is to be made to the posterior extremity of the turbinated bone, the instrument represented in Fig 7. is used.

Fig. 7.

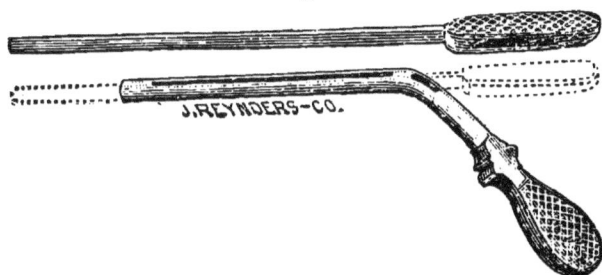

This is a modification of Dr. Andrew Smith's grooved
catheter for cauterization of nasal mucous membrane
by means of fuming nitric acid. This instrument is
smaller and much easier introduced, the canula is flat
and not round. Length of tube 6½ centi.; flat inside;
probe, extent 2 centi.; length of handle 5$\frac{8}{10}$ centi.; circum. of tube 1½ centi.; length of probe 9½ centi.
There is no need for *slot* on top of tube as represented in the drawing. It will be noticed that the probe
is 3 centi. longer than the tube or canula.

The probe is passed through the tube and around
its distal extremity a small piece of absorbent cotton
is twisted, and on one side (the side on which the application is to be made) the chromic acid paste is applied; the probe is then withdrawn into the tube;
the tube is now oiled with vaseline and gently pushed
into the nasal fossæ through the space between the
lower turbinated bone and the septum until the point
reaches the hypertrophied mass to be destroyed.
This can be ascertained by the touch or by posterior

rhinoscopic examination. After reaching this point, the tube is steadied by grasping the handle with thumb and fore-finger and placing the little finger (of same hand) on lip or cheek of patient and with the other hand the probe is pushed through the tube, then by depressing and elevating, the handle altern- ately, the entire mass may be thoroughly touched with the cautery. After accomplishing this, the probe is again drawn into the tube, and the instrument with- drawn. The cavity is now to be sprayed with Dobell's or some alkaline solution which relieves any pain caused by the application. By the use of this instru- ment we can avoid the touching of any part of the mucous membrane that we desire, the cautery being concealed, our application can be limited or general.

Where the mucous membrane is hypertrophied and pendulous, over the entire inferior turbinated bone, the application is made with instrument represented in Fig. 8. which is similar to Fig. 7. In this instru-

Fig. 8.

ment the tube is closed at its distal end and has slot on side.

Length of tube 9½ centi.; length of handle 6 centi.; circum. of tube 1½ centi.; length of slot 4½ mil.; length of probe 9 centi.

The tube is inserted into the nasal cavity with the slot to the side of the hypertrophy, the probe, cov- ered with absorbent cotton and the chromic acid, is then pushed quickly through the tube and the application made to the entire lower surface of the inferior turbinated bone. "The affinity of the acid for organic matter is such that it acts immediately. There is no pain of consequence resulting, and no bleeding. After the first application, our view of the remaining portion is not obscured by blood." Two or three applications is all that is necessary to remove the largest hypertrophy of mucous membrane that I have seen. I never apply the acid oftener than twice a week. After destruction of the membrane, it can be easily removed with small forceps or with loop of no. 5 piano wire attached to a probe. The after treatment consists in making applications on alternate days of the boric-acid-vaseline mixture before mentioned.

In addition to the treatment already suggested, a tonic is always given, the one generally employed by me, and with most gratifying results, is the Syr. of Hypophos. of Lime and Soda with Iron and Strychnia.

This preparation I have used largely for several years, and, as made by Mr. T. B. Wood, of Lexington, Ky., it has no equal, that I know of, as a general tonic. Each drachm (teaspoonful) contains 2 grs. of Hypophosphite of Lime and 1 gr. of Soda, 1 gr. of Phosphate of Iron and $\frac{1}{32}$ gr. of strychniæ. A teaspoonful is given three times a day.

For the itching and burning of the eyes, a Sol. of boric acid, 10 gr. to ℥j, is frequently used. For the rasping cough and headache, 5 gr. salicine, 3 gr. ammon. mur., ¼ gr. extr. belladonna, in capsule every 2 or 3 hours until relieved, has afforded good results. Sufferers from this disease are greatly annoyed by cold sweaty feet. Bathing the feet in salt water night

and morning, rubbing them dry, and thoroughly applying vaseline, soon relieves this.

Since the preparation of this paper in Oct., 1884, the
discovery of the magic effects of muriate of cocaine
has somewhat changed my views of the treatment of
Pruritic Catarrh. In the past three months I have
had five cases of the most aggravated variety of this
disease and have been able to give each one immediate relief by the use of a 4 *per cent.* Sol. of cocaine.
It was applied by means of absorbent cotton wrapped
around a thin piece of whale bone, about 2 inches
long, thoroughly saturated with the cocaine, gently
inserted into the nostril. The first application was
allowed to remain 8 or 10 minutes, when it was removed and a similar one made, using fresh cotton
and cocaine. The second application was allowed to
remain ten minutes longer. When it was removed and
the entire *lower border* of the inferior turbinated bone,
as well as the posterior surface of the inferior and
middle turbinated bones, were thoroughly cauterized
with chromic acid. Only one nasal cavity was treated
the same day. The after treatment has been the
same. There has been no unpleasant result following
this treatment, save a slight pain in the upper jaw
and the eye of the side to which the application was
made, which came on a few minutes after and lasted
from a half to one and a half hours. With the use of
cocaine I feel certain in promising my patients not
only temporary relief, but destroy the hypertrophied
tissue, in the acute stage, thus cutting short at once
the disease. For several days after this operation
there is a profuse discharge of mucus but no return
of the annoying hay-fever (?) symptoms, if the offending tissue is thoroughly removed.

When this operation is performed during the acute stage, I always advise rest of both mind and body for several days and make the application of boric-acid-vaseline with from 10 to 15 drops of the cocaine solution daily fora week.

I feel that I can endorse all the good that has been said of the local anæsthetic properties of cocaine. In an extended use of it in the nose and throat it has, in my hands, accomplished all that could be desired and with it we can now accomplish, what—a few months ago——would have taken weeks and months.

INDEX.

D.

Daly, Dr. Wm. H., xi,135, 143.
Dates of attack, table of, 27.
Deceptive sensations. 71, 88.
Defective methods of investigation. 19, 42.
Destruction of growths by Dr. Daly. 136.
Diagnosis and Prognosis. 100.
Diet. 122.
Differential table. 101.
Disappearance dates of. 27.
Disappearance, time of, in various stages. 101.
Diseased membrane, locating it. 139.
Diuretics. 116,144.
Donaldson, Dr. F. 154.
Drawers, cotton and woolen. 119.
Dust 97, 99.

E.

Earliest form. 85.
Effects of irritating treatment. 48, 49.
Effects of spray relieving. 114.
Elastic garters. 121.
Electrodes. 133.
Electricity. 115, 116.
Elliotson, Dr.33.
Eruption. 65.
Eucalyptus oil. 111, 143, 152.
Europe, a visit to. 96.
Eustachian tubes. 45, 69. 127, 151.
Exercise. 123.
Exhaustion from heat. 98.

Extracts from Dr. Daly's paper. 135.
Extravagent modes of expression misleading. 78.
Eyes. 65, 84.
Eyes affection of in various stages. 101.

F.

Faradic current. 115.
Fauces. 69.
Feathers. 122.
Flannel suits. 119.
Flashes of heat. 76.
Flowers. 98.
Flowers, the effect in various stages. 101.
First or formative stage. 100.
Forgetfulness. 47, 77, 80.
Fourth stage or form. 104.
" Fuzz " on the trees. 81.

G.

Galtheriæ oil. 113.
Galvano-cautery. 129, 130, 131.
Garters. 121.
Gin. 123.
Glycerinæ. 113, 152.
Gordon, Mr. W. 33.
Gream, Dr. 33.
Gums. 70.

H.

Hack. Dr. 143.
Hair. 117, 118.
Hair matress. 122.
Handkerchiefs. 123.
Hats. 119.